THE WORLD NATURALIST

Biological Clocks

Biological Clocks
Their Functions in Nature

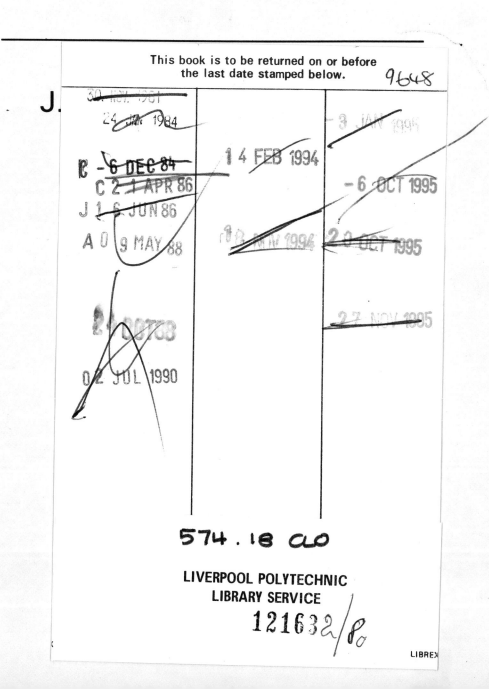

To my wife Anne

First published in Great Britain by
George Weidenfeld and Nicolson Ltd
91 Clapham High Street, London sw4

ISBN 0 297 77685 1

Printed in Great Britain by
The Camelot Press Limited

Contents

Plates

Figures

Introduction

DAY FOLLOWS night, new moon follows old, summer follows winter: 'So do flux and reflux – the rhythm of change – alternate and persist in everything under the sky,' wrote Thomas Hardy. Rhythmicity is indeed characteristic of many, though not all, natural phenomena. Shakespeare preferred to emphasize the passage of time rather than its cycles: 'When I do count the clock that tell the time, I see the brave day sunk in hideous night.'

A little reflection will reveal that the ticks of Shakespeare's clock, the linear aspect of time, may be thought of as a slow rate of change of its cyclic aspects: it is a witness that creeps along with leaden foot, and pauses not for man or beast. The Mephistopheles of Shakespeare's contemporary, Christopher Marlowe, knew this well:

> Stand still you ever-moving spheres of heaven,
>   that time may cease, and midnight never come . . .
> The stars move still, time runs, the clock will strike,
>   the devil will come and Faustus must be damn'd.

Most living organisms reflect their planetary origin in the possession of innate periodicities which are synchronized with the daily, lunar and seasonal changes that take place around them. Except for the bacteria and those algae that lack a discrete nucleus, probably all living organisms possess biological clocks by which they can measure the passage of time. In some cases, as we shall see, rhythmic activities result. In others, however, the clock may be used to time a single non-recurring event in the life of its owner. 'The flowers anew, returning seasons bring!/But faded beauty has no second spring,' wrote the Romantic English poet Ambrose Philips.

This volume is neither a text-book nor a general treatise on rhythms and cycles in nature; it is a consideration of their adaptive significance, and the ways in which physiological clocks are made use of by living organisms, including man. Nevertheless, it should be remembered that the topic is one whose biological importance has but recently become generally manifest; it is also a subject of great intrinsic beauty.

In the objective study of rhythm the student of nature is simultaneously daunted and exalted. In Edward Fitzgerald's translation of *The Rubáiyát of Omar Kayyám* we find:

> Ah, fill the Cup:–what boots it to repeat
> How time is slipping underneath our Feet:
> Unborn TO-MORROW and dead YESTERDAY,
> Why fret about them if TO-DAY be sweet!

Let us, likewise, study biological rhythms and enjoy the study. Although 'The Flower that once hath blown for ever dies,' its reproductive activity, controlled by an internal clock, generates the seeds that will survive to produce next year's blossoms. This point of view surely justifies some optimism. The life of today holds the seeds of tomorrow, and who can be certain that it holds no more?

Aesthetic delights arise from causes that are essentially rhythmic–light and colour, music and poetry; for the essence of beauty lies in rhythm and harmony. The study of biological rhythm involves philosophical as well as physiological and ecological considerations. Although an objective, scientific attitude has been adopted as far as possible in the following pages, it should be remembered that considerations of a different kind may be lurking not so far beneath the surface.

My object is to give an account of the functions of biological clocks and of the rhythms they engender in relation to environmental cycles. (A glossary of terms used is provided at the end of this book.) From a discussion of the living clocks of nature, and a consideration of the ways in which they are synchronized with the physical and meteorological cycles experienced on the surface of the earth, we turn to the control of flowering in plants by the measurement of day-length or photoperiodism, the importance of day and night in the lives of animals, and the advantages to an organism of being preadapted to forthcoming changes. Rhythms of feeding, predation and dispersal are controlled by biological clocks, probably the very same cellular chronometers that are employed in celestial navigation.

The rhythm of day and night is not the only one to influence the lives of plants and animals. Other major cycles include those of the moon and tide, and the annual progress of the seasons which is reflected in phenomena such as hibernation, reproduction and migration; these cycles are controlled by the interactions of a yearly chronometer with the measurement of photoperiodism by the daily clock.

Indeed, the time of life is, to a great extent, the cyclic time of nature. As Joseph Addison observed:

> Th' unwearied Sun from day to day
> Does his Creator's power display ...
> Soon as the evening shades prevail
> The Moon takes up the wondrous tale,
> And nightly to the listening Earth,
> Repeats the story of her birth.

Acknowledgement

My thanks are due to Dr J. T. Fraser for constructive comments on the manuscript.

Chapter 1

Living Clocks

Over 2,400 years ago, Hippocrates advised his associates that regularity was a sign of health, and that irregular body functions promoted an unsalutary condition. He counselled them to pay close attention to fluctuations in their symptoms and to look at both good and bad days in their patients and in healthy people. Around 300 BC, Herophilus of Alexandria is said to have measured biological periodicity by timing the human pulse with the aid of a water clock. Early Greek therapy involved cycles of treatment known as 'metasyncrasis' during which patients were not fed the same food and herbs, or given the same exercise every day.[1]

Since the observations of Aristotle some twenty-four centuries ago, it has been known that the ovaries of sea-urchins acquire a greater size than usual at the time of full moon. During the march of Alexander the Great into India during the fourth century BC, Androsthenes, a Macedonian philosopher who accompanied him, noted daily movements of the leaves of the tropical tamarind tree. The Roman orator and statesman, Cicero, mentioned that the flesh of oysters and all shell-fish increased and decreased with the moon, while according to Pliny the Elder (AD 23–79) careful observers attributed to lunar power the rhythmic increase and decrease that occurred in the bodies of oysters and shell-fish. In fact, lunar rhythm in living organisms – which we now regard as the manifestations of an internal, biological clock with lunar periodicity – impressed the ancient writers rather more than have diurnal rhythms. Perhaps lunar rhythms are less obvious, hence their significance appears more mystical, and before electric lighting, moonlight was very important.

Biologists have long been interested in the ability of plants and animals to adapt themselves to the physical conditions of their environments; but the significance to them of the cycles of daylight and darkness, of the alternation of spring and neap tides, of the phases of the moon and of seasonal changes has been overlooked until comparatively recently. Thirty years ago, when I first became interested in the subject, the number of biologists working on rhythmic phenomena throughout the world cannot have exceeded a score. Today, more than a thousand research papers on biological rhythms are published annually. Even so, most of these are concerned with elaborating the physiological mechanisms of biological clocks and only a few attempt to analyse their ecological significance or assess the adaptive advantages of being able to measure the passage of time.

Rhythmic activity of one kind or another has now been described in cells, organs, organisms and populations. It is manifest in phenomena as diverse as the timing of luminescence in microscopic marine dinoflagellates (unicellular organisms), and the control of fluctuations in human body temperatures.[2] Most plants and animals possess innate periodicities which are synchronized with the daily, lunar and seasonal changes that take place in their natural environments. There may be no selective advantage in measuring time to organisms, such as bacteria, whose life-span is typically less than a day but, to plants and animals that live far longer than twenty-four hours, there must be clear advantages in being able to anticipate the cyclical changes which occur during that time. These advantages, and those which accrue from the measurement of longer intervals, will be analysed in later chapters.

The clock mechanism

Few biologists today doubt that organisms possess biological clocks, but there is a degree of controversy regarding the fundamental mechanism involved. Thus, for over thirty years, F. A. Brown Jr and his co-workers have maintained that the clocks of living organisms are timed by subtle, rhythmic, geophysical forces.[3] Most other investigators, however, now hold that biological rhythms are endogenous (that is, internal in their control), and completely independent of the environment for their fundamental timing.[4] Insofar as all biological clocks are part and parcel of their evolving rhythmic environment a good case can be made in support of the idea that biological clocks are both endogenous and exogenous, but to different extents. Although the physiological clock may be endogenous in the usually accepted sense, it cannot be divorced from its genetical ancestry which has been influenced by natural selection.

Thus the rhythms of plants and animals must have arisen, in an evolutionary context, from cellular phenomena which have subsequently been strengthened by natural selection. Under experimental conditions they exhibit certain fundamental qualities that are common to them all. For example, the phase of a twenty-four-hour rhythm – that is, the position in the cycle at which some particular event takes place – is not necessarily restricted to any particular time of night or day. Secondly, although biological phenomena are usually extremely sensitive to thermal influences, the periods of biological rhythms are relatively independent of changes in the ambient temperature. If this were not so, of course, biological clocks could only function in constant environments. At the same time, although the length of the *period* of a rhythm may be relatively temperature-independent (above the minimum threshold below which activity is suppressed), nevertheless its amplitude will vary according to the biological temperature coefficient of the process and it may be synchronized by temperature changes. Some physiological processes are more affected than others by changes of temperature. Similarly, although the amplitude of a rhythm is sensitive to metabolic inhibitors, such as sodium cyanide and other narcotizing agents, investigations have shown that the period is normally unaffected. Finally, rhythms are not normally learnt, because apparently arhythmic organisms, raised under constant conditions in the laboratory, may become rhythmic after experiencing only a single non-periodic stimulus, such as a flash of light.[4] As we shall see later, however, there are sensitive periods in the learning of animals.

Biological clocks are thought to be self-sustained oscillations whose phase can be reset or entrained, according to Colin Pittendrigh and V. G. Bruce, by an external synchronizer or *Zeitgeber*, a term proposed by Jürgen Aschoff.[5-8] In their natural habitats, most plants and animals display nycthemeral rhythms which last for a period of a day and a night, and are entrained to a frequency of twenty-four hours by the daily cycles of light and darkness engendered by the rotation of the earth on its axis. Although these rhythms frequently persist under constant laboratory conditions, the free-running periods usually become either slightly longer or somewhat shorter than twenty-four hours. For this reason, Franz Halberg has coined the word 'circadian' (from the Latin *circa*, about, and *dies*, a day) to describe them[7]. Persistent lunar and tidal rhythms are likewise called 'circalunadian' (about a lunar day) or 'circalunar', because the period of the bimodal lunar-day rhythm is usually longer or shorter than the

7

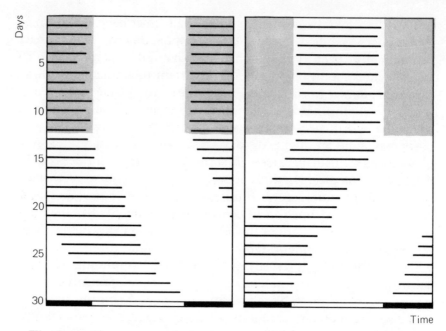

1. The times of locomotory activity of a generalized nocturnal animal (e.g. a mammal), left, and a generalized day-active animal (e.g. a lizard or bird), right, during a 12-day sojourn in a 24-hour light–dark regime followed by 17 days in constant illumination during which the organism typically adapts a free-running period either longer, left, or shorter, right, than 24 hours. Shaded areas indicate darkness. Horizontal lines show times of activity. The white strip indicates the light period from 06.00 to 18.00 hours. (After F. A. Brown Jr.)

period displayed in nature, while endogenous yearly rhythms are called 'circannual' for similar reasons.[9, 10]

Under certain conditions, oscillations can be entrained to show periods which are a multiple of the entraining cycle. This phenomenon is called frequency multiplication, since the entrained period which the rhythm shows is longer than that of the external cycle. Frequency demultiplication – transformation of the output from an oscillator (the circadian clock) by counting off a certain number of beats – may be the explanation not only of circalunar rhythms but also of oestrous cycles and other long-term periodicities, such as those revealed in man by the study of psychiatry and medicine.[11] Frequency demultiplication occurs in nerve fibres; it has also been advocated as one possible explanation for the accuracy and temperature independence of circadian rhythms. The latter are thus supposed to be generated through frequency demultiplication from high-frequency oscillators on the molecular level. Although the application of cybernetic principles has sometimes achieved satisfactory physical models which suggest working hypotheses, it would be a grave error to pursue the analogy too closely.[12]

With regard to circannual clocks, only a handful of biological rhythms have been shown to persist with a period of approximately 365 days, under conditions held constant with respect to light and temperature, for at least two or three years. Examples are provided by hibernation in ground squirrels and the reproduction of European starlings, cave crayfishes and marine coelenterates. Even in such relatively well-documented cases, however, twenty-four-hour light cycles are commonly used and are considered to be constant. They could theoretically be summated or added to one another, as they may also be in the case of lunar cycles–though it would be hard to estimate whether frequency demultiplication of a circadian oscillator, an endogenous circannual clock, or extrinsic timing, provides the most plausible explanation of the phenomenon. It is, moreover, difficult to imagine the adaptive significance of circannual rhythms since, in most parts of the world, changes in photoperiod, measured by the circadian clock, provide a reference point for every phase of the calendar year.[13, 14]

In higher animals, the circadian clock is responsible for triggering a wide range of responses, both at cellular and multicellular levels of organization. Throughout the day, whole series of activities are generated in different parts of the body, and numbers of different clock centres have been identified, such as the mammalian hypothalamus and the pineal gland of birds. As well as central clocks, there is an hierarchy of lesser clocks, some of which are demonstrably independent of one another. Whether the same mechanism is common to all cellular clocks, or whether convergent evolution has produced several different types of cellular clocks, is not yet known. Indeed, it is often difficult to determine whether a particular function, such as enzyme activity, engenders the circadian oscillation, or is being driven by it.

Although the ability to measure time is apparently an innate property of the cell, the fundamental nature of biological clocks remains obscure. Feedback regulation of enzyme activation and inhibition; RNA (ribose nucleic acid) synthesis; ionic diffusion across cellular membranes; and a chronon model which proposes a circadian transcription of genes (the rhythmic transfer of genetic information in DNA into an intermediate RNA through which it is then translated into proteins),[15] have all been proposed as possible mechanisms.

9

Synchronization

Light is the most usual and important synchronizer of circadian rhythms but temperature, too, may occasionally be effective. Circadian rhythms, as previously stated, can sometimes be initiated in arhythmic organisms under non-inhibitory conditions by a single stimulus, such as a flash of light or a brief increase or decrease in the ambient temperature. In other organisms a series of repeated stimuli may be required before a stable phase relationship has been reached.[16] In nature, thermal influences probably reinforce or supplement those of light intensity. Light and temperature are the only environmental factors so far conclusively demonstrated to be coupled to the circadian biological clock, but it is by no means improbable that some other regularly repeated stimuli, such as periodic noises, social cues, or changes in barometric pressure may also be effective phasing agents.[17-22] Of course, if Frank Brown's hypothesis of exclusive exogenous control of biological clocks is correct, then we shall have to add to this list all active geophysical forces, whatever they might be. In higher animals and man, the number of possible synchronizers is larger because of the many important correlations to the environment. Thus it is often difficult to demonstrate the endogenous rhythm as clearly in them as it is in plants and lower animals.[23]

Some biological activities are non-rhythmic, and others can be entrained only with difficulty. The significance of this will be discussed later. A number of attempts have been made to entrain biological clocks to periods of other than twenty-four hours, but they have not been very successful. In most cases the imposed cycles have either failed to persist under constant conditions even though entrainment was successful, or else have merely reinforced the twenty-four-hour clock.[18, 24] They may be the result of learning.[25, 26] The interactions between light and temperature have also been studied on a number of occasions. Temperature cycles have been shown to dominate the rhythm of petal movement in *Kalanchoë blossfeldiana*, a succulent orpine plant, while a light–dark cycle entrains the rhythm of eclosion – the emergence of the adult insect from the pupa or chrysalis–in fruit-flies (*Drosophila pseudoobscura*) more strongly than do temperature fluctuations. No doubt the relative influence of light–dark and temperature cyles depends on the magnitude of the change in temperature and the incident radiant flux employed.[26]

Environmental synchronizers are not equally effective at all hours for, as Erwin Bünning has been emphasizing for decades, the sensitivity of an organism to its *Zeitgeber* fluctuates rhythmically. Bünning began with two postulates: first, that there is a circadian rhythm of some physiological function of plants that is vitally associated with flower induction in both short-day plants and long-day plants; and second, that plants make use of this rhythm for measuring time. The timing mechanism that is responsible for the photoperiodic response of flowering is the same as the mechanism that is responsible for the timing of leaf movements. This timing mechanism has two alternating phases of about twelve hours each, which may be distinguished as photophil (light-loving), and scotophil (dark-loving). Photophil is equivalent to the day phases and scotophil the night phase of a circadian rhythm. Consequently, light falling on a plant during the photophil phase will enhance flowering but, during the scotophil phase, will inhibit it.[27, 28]

In a similar way, light shocks of ten-minute duration given to flying squirrels (*Glaucomys volans*) which are otherwise maintained in constant darkness, shift the phases of their activity cycles only if presented during the animals' subjective night.[29] However, such variations in the reactions of an organism to perturbations could equally well be due to the properties of its cellular clocks, the clock's hands, their mediating pathways or a combination of all these components.[22]

Exogenous and endogenous rhythms

An early worker in the field of biological rhythms, Orlando Park, subdivided rhythmic phenomena into two main categories: exogenous rhythms, which are a direct response to physical change in the environment and do not persist when conditions are kept constant, and endogenous rhythms which continue, at any rate for a time, under constant conditions.[30] Park suggested that the majority of species, in their natural environments, appear to show a combination of both types, and such rhythms he designated composite. A composite rhythm, as thus defined, would differ from some endogenous rhythms only in that it would become entrained instantly. Since the rapidity of synchronization may depend upon the intensity of the *Zeitgeber*, there can be no clear distinction between composite and endogenous rhythms.

Many animal rhythms which appear to be exogenous do not in fact persist under constant environmental conditions. Nevertheless, they cannot readily be induced by environmental cycles having a periodicity differing markedly from twenty-four hours. For this reason, unequivocal examples of true exogenous rhythms, especially of locomotion, are rare. Probably most rhythms which are apparently exogenous actually represent the expression of endogenous cellular clocks that rapidly get out of phase with one another when removed from the influence of environmental periodicity.[21]

Philip Corbet distinguishes between endogenous rhythms, and cycles. He defines a cycle as 'a sequence of events, repeated during the life of an individual, and characterized by a change in physiological condition or behaviour. The events always fall in the same order, but the time intervals between corresponding ones are not necessarily constant. Cycles of feeding and ovarian maturation in blood-sucking Diptera fall into this category.'[31] The separation of rhythms from cycles may, however, cause difficulty. Cyclic events, such as oviposition cycles, naturally tend to become regular in time under constant conditions. For instance, green-bottle flies (*Lucilia sericata*) lay batches of eggs at intervals of approximately three days in the presence of meat. At other times they will only lay three days after meat has been presented; if meat is given every fourth day, a four-day oviposition rhythm can be produced. Should this be regarded as an exogenous rhythm, induced in a cyclic but non-rhythmic event by a regularly appearing stimulus; or as an endogenous rhythm whose phase has been delayed by the absence of the appropriate 'permission'?[32]

The reproduction of certain desert birds is engendered by a combination of two factors – their innate annual reproductive rhythm and the stimulus of rainfall.[33] The latter provides the necessary 'permission' to lay, for oviposition is inhibited without it. In other words, an endogenous rhythm must represent the manifestation of a biological clock, but a cyclical event may show exogenous periodicity in the presence of a regularly appearing stimulus (for example alternating light and darkness) and 'permission' (for example temperature above a threshold value). Centipedes (*Scolopendra* spp.) show a marked endogenous rhythm in constant darkness, but this is suppressed by constant light. By analogy, the oviposition rhythm of green-bottle flies can be regarded as being suppressed by the absence of meat. Alternatively, as suggested above, the presence of meat may be looked upon as 'permission' for a cyclical event to be manifested regularly. Clearly the two concepts coalesce at this point.[32]

Rhythmic changes in the physical factors of the environment may not only synchronize the phases of biological clocks with the time of day; they may also elicit immediate responses from the organism. Whether an observed periodicity is regarded as exogenous or endogenous depends upon its subsequent persistence under constant conditions. A problem therefore arises as to whether an observed periodicity which disappears rapidly under constant conditions represents a true free-running circadian rhythm, as I have suggested, or an exogenous periodicity that has persisted for a short while under its own inertia and is not driven by a biological clock. To use an analogy, if the wagons of a railway train are uncoupled in motion, they may follow the engine for a while before coming to a halt. The nycthemeral activity rhythm of the desert locust (*Schistocerca gregaria*) is not clearly marked in nature. It could theoretically represent an exogenous periodicity that persists briefly in constant conditions but is not controlled by a biological clock. Although activity of the desert locust is greater in constant light than in darkness, its periodicity persists only in darkness. To pursue our analogy further, the second condition (darkness) can be likened to uncoupling the wagons while the train is running downhill, the former (constant light) to uncoupling them when the engine is pulling them up a slope. The existence of an underlying biological clock can, in some cases, be revealed only by the reluctance of a rhythm that is apparently exogenous to become entrained to any periodicity other than one of about twenty-four hours.[32] This applies to the desert locust which in fact possesses an excellent clock to which it pays scant heed under natural conditions. The same is true of many human physiological rhythms which may only appear as symptoms of disease.

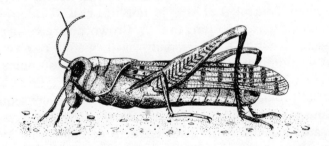

2. The desert locust must be able to be active at any hour of the day or night in order to exploit its inconstant environment. (From J. L. Cloudsley-Thompson.)

Effects of constant conditions

When circadian biological clocks are free-running under constant conditions, as we have seen, they seldom maintain periods of exactly twenty-four hours. For instance, although bullfinches kept in darkness show a period of about twenty-four hours' duration, this changes to twenty-two hours in constant light. In rodents, on the other hand, the period of spontaneous locomotory activity is longer in light than in darkness.[8, 34] With illumination, the daily period of activity becomes progressively later on successive days, and may be shifted steadily round the clock, with no tendency to be fixed at any particular hour of the solar day or night. In general, it seems that the activity rhythms of nocturnal animals are delayed by constant light, while those of day-active animals tend to be accelerated. The extent to which this occurs depends upon the intensity of the light.[22] Although a number of exceptions are known to this circadian rule, it may be significant that many of them are provided by tropical species in whose natural environments the length of daylight varies little throughout the year.

When it does occur, however, this shift in phase allows the daily rhythm of an animal to keep pace with the seasons as the days lengthen in spring or draw in during the autumn; for the duration of daylight in temperate regions varies significantly according to the time of year. Obviously, if an animal is to maintain its regular daily activity it cannot synchronize both to dawn and to dusk, since the period between them is variable. It seems, however, that most nocturnal animals tend to use dusk as the synchronizer, while day-active species use the dawn.[22] Aschoff 1960.

It might be thought that the existence of free-running, circadian periodicity would conclusively invalidate F. A. Brown's hypothesis of exogenous control, but this is not the case. Under so-called constant conditions, according to F. A. Brown, an organism reaching a light-sensitive phase in its daily cycle would receive from a constantly illuminated environment a shifting stimulus whose strength would be a function of the light intensity. Thus, constant light would provide rhythmic synchronization as a consequence of the organism's own responsiveness.[3] Brown calls this hypothetical process autophasing. It is instructive to compare this idea with Bünning's hypothesis of entrainment through light-sensitive phases of the circadian rhythm.[25]

We have seen that temperature cycles can be used to synchronize the phases of biological clocks, even though they are usually much less effective than are cycles of light and darkness. At the same time, the periods of circadian clocks are relatively independent of temperature: in general there are only slight, although quite significant, differences between the steady-state periods of a circadian rhythm at different ambient temperatures. On the other hand, the amplitude of any particular rhythmic process may depend greatly on the ambient temperature. When organisms are cooled down below a threshold – usually between zero and 10°C – their clocks are stopped. A persistent phase-shift is later seen after they have been transferred back to normal temperatures; this shows that the clock itself has been stopped and not merely the processes controlled by the clock. This effect has been described in the movements of bean leaves, plant growth rhythms, the time-sense of bees, pigment migration in fiddler crabs, and the locomotory activity of insects and spiders.[25, 35]

The unicellular clock and its environment

Without doubt, the most far-reaching research on this subject has been carried out by J. W. Hastings and B. M. Sweeney, who have investigated circadian rhythms in certain marine dinoflagellates (*Gonyaulax polyedra*). These authors studied four different rhythms: a flashing luminescence rhythm obtained when the cells are stimulated mechanically; a rhythm in spontaneous luminescent glow; a rhythm in the photosynthetic capacity of the cells; and a rhythm in cell division. These rhythms show circadian periodicity and can be inhibited independently of one another but are resumed in phase after removal of the inhibitor. At the same time, a relatively brief exposure to light can cause the rhythms to be shifted by several hours.[4, 36]

As in the rhythms of more advanced organisms, the endogenous circadian rhythm of stimulated flashing in dinoflagellate populations can be initiated by a single abrupt change in illumination. Individual isolated cells behave as the population does, losing all overt rhythmicity in constant bright light, and recovering it when transferred to light of lower intensity. A similar loss of rhythm occurs in constant light at temperatures below 13°C. When the temperature is increased the rhythm is re-initiated and the phase determined. Thus the biological clocks of dinoflagellates can be synchronized or have their phases shifted by changes in light and temperature. Circadian rhythms have been demonstrated in various functions of

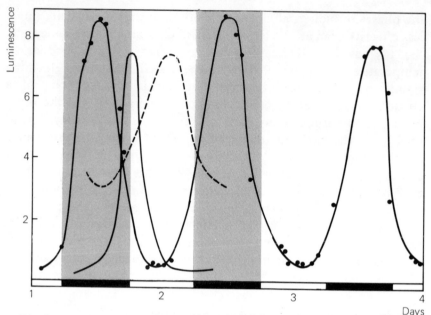

3. Rhythms in *Gonyaulax polyedra*. At left are shown several rhythms and their phase relationship in a daily light–dark cycle. The dashed curve represents photosynthesis as measured by the rate of ^{14}C uptake; the single acute peak represents the biological glow rhythm; and the third curve, the stimulated bioluminescence rhythm (cells are induced to flash by mechanical agitation). All three rhythms will persist in constant conditions (although only the persistence of the latter is portrayed in this figure). Note that the period of this rhythm is not 24 hours, as it is in the light–dark cycles, but becomes longer (24.2 hours) in constant dim illumination. The three rhythms are not in phase with one another in natural illumination cycles. Luminescence is assayed in terms of the total amount of light emitted when cells are stimulated to exhaustion. The black strips represent 12-hour periods from 18.00 to 06.00 hours. (After J. W. Hastings.)

other unicellular organisms, such as phototactic sensitivity and swimming in flagellates (*Euglena gracilis*), sexual reactivity in the ciliates (*Paramecium bursaria*) and so on. The chronon concept of the cellular clock is based on such work.[15] Since all the circadian rhythms of a unicellular organism appear in phase (synchrony) and never in allochrony, it must be assumed that they are coupled to a single pacemaker.

Interactions between cellular clocks within multicellular organisms

Since every eucaryote cell (a cell in which there is a membrane separating the cytoplasm from the nucleus) contains or is itself a

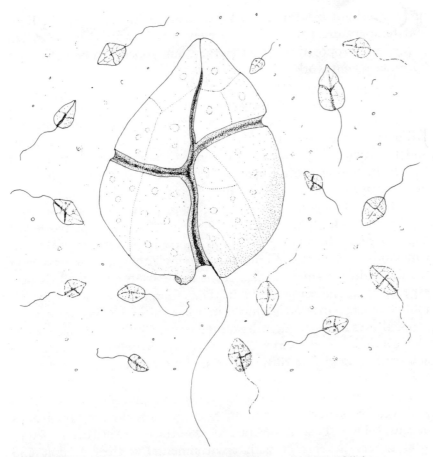

4. *Gonyaulax polyedra*, a marine dinoflagellate whose rhythms of light production and photosynthesis have been studied in much detail.

clock, the question arises as to whether, in the multicellular organism, the cells are synchronized with one another directly or only via the operation of some exogenous *Zeitgeber*. The evidence seems to suggest that both conditions occur, especially as some rhythms appear synchronously, and others allochronously.

Many years ago I found that field crickets (*Gryllus campestris*) are normally active during the daytime, the rhythm being endogenous and its frequency temperature-independent. When the circadian rhythm has died away after several weeks in constant conditions it can be re-established by a single exposure to light, or by a return to higher temperatures after a period at 5°C. The observation was interpreted as implying that the loss of rhythm was due to the fact that the various cellular clocks of the insect had become out of phase with one another under constant conditions, and were re-synchronized by the light and temperature shocks.[37] This view was in accordance with an hypothesis proposed by Janet Harker, and appears to be supported by more recent work.[18, 38]

The nuclei of syncytia (masses of cytoplasm, enclosed in a single continuous plasma membrane, containing many nuclei), or of multinucleate cells, nearly always divide in perfect unison, but ordinary cells seldom show such an effect. Cell division or mitosis and the meioses (reduction divisions) of animal testes and plant anthers nevertheless often exhibit a very high degree of synchrony, as do the early cleavage stages of developing eggs.

Although unusual, synchrony of mitotic rhythms is also well known in plants, particularly in algae and in the root tissues of angiosperms. The phase may vary in different species, however, with a maximum frequency of mitosis occurring either at night or during the day. Mitotic rhythms also occur in a variety of animal tissues, especially the epidermis, and it has been found that a cold shock may induce synchronization of mitosis in tissue cultures.[39] In such cases it is not always clear whether the timing of the rhythm is exogenous or endogenous, but the latter is more likely. Synchrony is probably achieved through the diffusion of metabolites associated with the synthesis of DNA (see page 9).

There may be no need for a central clock if all cells possess clocks which can be synchronized in this way. Nevertheless certain groups of cells, such as those in the optic lobes of the brain, are better placed than others to entrain with environmental cycles of light and darkness. These might, therefore, logically be expected to become specialized as master clocks, although few have yet been positively identified.

A cell can influence other cells by what it removes from, or discharges into, the common aqueous medium. The ability of many cells to act in consonance depends on the restraining influences they have on one another. It is difficult to conceive of mechanisms, other than metabolic or hormonal ones, by which individual cells can synchronize with one another directly. The responses of an organism to changes in its environment are in the final analysis, the result of the responses of individual cells to changes in their local environments.[40, 41] Similarly, the overt rhythm of an organism represents the sum of the reactions of its constituent cells to their biological clocks which, in turn, may be synchronized by a group or groups of cellular clocks forming a master clock.[21, 42] In general, the master clocks which have so far been studied in animals appear to fall into one of two groups: those which stop in continuous light, such as the eclosion clocks of fruit-flies[5, 17, 43] and other insects, and those involved with activity rhythms.[44] Clocks of the first type seem to be associated mainly with developmental rhythms and to be under neuroendocrine control; those of the second are the direct product of a complex nervous system. Their phase-response curves, in general, are more stable against interrupting light perturbations since the maximum phase-shifts are usually small. It is, at present, not known by which type of clock the time-sense and photoperiodic responses of animals are controlled.

Circadian clocks of the second type have been implicated in many diverse kinds of behaviour. Generalized activities, such as locomotion, are influenced by an array of factors so complex that any clock system must interact with a network of other internal and external stimuli. In many instances the coupling between the master clock and effector organs may be hormonal, for example testosterone can de-synchronize the circadian rhythms of birds and mammals. For fifteen years it was accepted that the circadian rhythms of insects were under the control of a hormone secreted by the suboesophageal ganglion.[45] Subsequent research having failed to confirm the central importance of this ganglion, it now seems more probable that hormonal rhythms are not primary driving oscillators but are driven.[46] Indeed while the physiology of the oscillator mechanism (which represents the escapement of the biological clock) is not yet understood, its coupling now appears to be primarily nervous (electrical) and only secondarily hormonal.

This line of thought is supported by the following reasoning. The neurosecretory system plays a central role in mediating between environmental changes and the responses of the individual. Neurosecretory cells have a capacity for receiving messages from the nervous system and for controlling the activity of other endocrine organs. Thus internal physiological processes are closely related to external environmental events. The adaptive advantages of this form of integration are evident: avian reproductive cycles are geared to coincide with environmental conditions favouring survival of the offspring, metamorphosis occurs in many insects when food plants are most abundant, and so on. By its very nature, the neurosecretory system is able to link internal responses with external stimuli, and is present in such phylogenetically diverse groups as the Coelenterata, Arthropoda and Vertebrata. In the more primitive phyla neurosecretory structures are situated diffusely and act directly to control processes such as growth and reproduction. During the course of evolution they have become more localized and concentrated in specific areas of the central nervous system where they are directly influenced by the brain and whence their effects are mediated through epithelial endocrine glands.[46]

Environmental synchronization of circadian rhythms

The entrainment of circadian rhythms with daily environmental cycles is achieved mainly through changes in light intensity at dusk and dawn, enhanced by the influence of the circadian rule.[16] In addition, synchronization may be assisted by those changes in ambient temperature which normally occur at these times. This is suggested by the fact that when the activity rhythms of various insects, including spider-beetles (*Ptinus tectus*) have been gradually lost in constant conditions, they can sometimes be re-initiated by periodic exposure to high and low temperatures. Regular thermal fluctuations also appear to prevent loss of the activity rhythm of the cockroach (*Periplaneta americana*) in constant darkness, although they have no power to initiate a rhythm once it has died out. The effects of temperature perturbations depend on the phase at which they are administered; they are greatest when operating in co-operation with the cycle of daylight and darkness.[35] Changes in relative humidity, barometric pressure, food supply and other environmental variables may sometimes support or inhibit the primary *Zeitgeber* (light) and its chief subordinate (temperature).

Physiological processes often result from the interactions between endogenous rhythms and external periodic synchronizers, especially changes in light intensity and day length. Synchronizers are sometimes necessary for the maintenance of rhythms as well as for their synchronization with environmental periodicities.

In order to be effective a synchronizer must be reliable. No doubt this is why biological clocks are synchronized by light more often, and to a greater extent, than by any alternative environmental variable, even though other factors such as temperature, relative humidity or predation may, at times, have greater ecological significance.[47] John Baker has distinguished between two groups of factors which influence the reproductive cycles of higher vertebrates: 'proximate' factors, in response to which the reproductive organs undergo their physiological development from a seasonally quiescent state to a functional breeding condition; and 'ultimate' factors which exert a selective pressure, ensuring that animals breed at the optimal season.[48] In many animals the most important ultimate factor is the availability of suitable food supplies for the young. Other selective influences, however, impose numerous compromises and adaptations on this basic principle.

From the above account it will be seen that all the complex harmony of natural rhythm and counterpoint could be based on the operation of simple metronomes – cellular circadian clocks – coupled directly and indirectly with one another, and synchronized primarily by the natural periodicity of daylight and darkness. At the same time, the circalunadian clock is probably coupled to these circadian clocks in some complex manner while, by using them to measure photo-period, seasonal cycles of reproduction, diapause and other physiological processes are synchronized with environmental influences resulting from the movement of the earth in solar orbit.[49]

Biological clocks have evolved through the influence of natural selection, and this selection is responsible for their efficiency. But what are the uses of efficient clocks, and why should living organisms have evolved them at all? Sunrise can tell them when it is morning, while sunset heralds the approach of darkness. Why, then, should they benefit by being able to measure the passage of time? This is the all-important question to which we shall address ourselves in the chapters that follow.

Chapter 2

Photoperiodism and Flowering

ALL LIVING organisms are able to measure the passage of time and express this capability in their endogenous rhythms. Not surprisingly, therefore, rhythmic phenomena, in both plants and animals, have a wide frequency spectrum. Physiological rhythms, such as enzyme-substrate systems and rhythms of respiration, excretion or heart-beat in animals, or of leaf or stomatal movement in plants, tend to show rapid oscillation and to be comparatively independent of external stimuli. Such elementary rhythms may, to some extent, depend upon temperature – for example, our own heart-rate increases when we have a fever – yet they are not entrained by cyclical processes occurring in the environment, nor do the lengths of their periods have any ecological significance. Cycles of daily, lunar or seasonal frequency, on the other hand, ensure the preadaptation of plants and animals to the fluctuations of environmental factors. The functions of such rhythms, therefore, are mainly ecological and are of primary concern in the context of the present volume.

Circadian rhythms in plants

It was a French astronomer, Jean de Mairan, who, in 1729, first realized that such movements were not merely under the control of environmental factors but were, in fact, regulated from within, for they continued when plants were enclosed in dark places.[1] Many of the processes of living plants have since been shown to exhibit such circadian oscillations. The most widely investigated of these have been the movements of leaves, often known as sleep movements, a description coined by Carl Linnaeus in 1751. De Mairan's findings were subsequently confirmed by Henri-Louis Duhamel de Monceau in 1758, and by J. G. Zinn the following year,[2] while, in 1832, Augustin P. de Candolle found not only that the leaf movements of the sensitive plant, *Mimosa pudica*, persisted in artificial light, but that their period was then shortened by about an hour and a half.[3] No particular importance was attached to this observation at the time, although it is now clear that it provides an example of a free-running biological clock accelerated by constant light, as discussed in the previous chapter. De Candolle also looked at plants maintained in darkness and observed that movements persisted, but less regularly than in continuous light. He found that many other species of plants showed sleep movements, including the leaves of the wood sorrel (*Oxalis acetosella*) and of the scarlet runner-bean (*Phaseolus multiflorus*). He also discovered that it was possible to reverse the course of leaf movements by placing the plants in darkness during the day and illuminating them at night.

During the second half of the nineteenth century many eminent biologists, Charles Darwin among them, worked on the patterns of plant movements.[4] There followed a period of stagnation until the nineteen-thirties, when the investigations of Erwin Bünning and his co-workers heralded a new era in rhythm research.[5]

Movements of the leaves of the scarlet runner-bean and other Leguminosae, as well as those of the telegraph plant (*Desmodium gyrans*) and the fronds of certain ferns (*Marsilia* spp.), are caused by changes of turgor pressure within specialized groups of cells at the bases of the leaf stalks. Furthermore many flowers open by day and close at night; light is responsible for the opening of the daisy (*Bellis perennis*) and of the wood sorrel, but it causes the flowers of the evening primrose (*Oenothera biennis*) and of tobacco plants

5. Bean seedling leaves in day (left) and night (right) positions, illustrating the circadian rhythm.

(*Nicotiana tabacum*) to close. The leaves of the wild balsam or jewelweed (*Impatiens aurea*) also droop under the influence of light. Flower movements are probably due to unequal growth on the two sides of the petals; this phenomenon is called 'photonasty.' Other circadian plant rhythms include movements of the petals of *Kalanchoë blossfeldiana*; exudations from the root systems of severed sunflower plants; odour production in *Cestrum nocturnum*; phototaxis in diatoms; and spore discharge from fungi (*Neurospora* and *Pilobolus* spp.).[6]

Unicellular organisms, such as *Gonyaulax polyedra*, as well as certain green algae (*Acetabularia* spp.) show rhythms in photosynthesis. Among higher plants, coffee (*Coffea* spp.) and several succulents exhibit a rhythm in their carbon dioxide compensation point which persists in continuous light. Succulents also show rhythms of carbon dioxide output in continuous darkness and at low light intensities. Rhythms of metabolism in higher plants are closely associated with the rhythms of stomatal movement which, of course, are important in regulating the exchange of carbon dioxide and oxygen.

It is clearly advantageous for plants, as it is for animals, to adapt their biological functions to the environmental conditions of the solar day and night. Photosynthesis can only take place in daylight, so both leaf movements and metabolic rhythms may perhaps be adjusted to this. If such be the case, however, it is difficult to understand why leaf movements should not be more widespread throughout the plant kingdom. Again if the movements of heliotropic flowers, which continually face towards the sun, have the adaptive function of providing maximum warmth for the developing seeds and thus of speeding their maturation, why do most other flowers not do likewise? It is clearly economical that flowers should be open only at times when their insect pollinators are active and thus not waste their fragrance but, doubtless, other factors are also involved.

Photoperiodic control of flowering

The most important function of the endogenous rhythms of plants is to measure time, which is essential for photoperiodic control. Plants living at latitudes some distance from the equator respond in their flowering to seasonal changes in the length of the day.

The influence of the relative lengths of day and night on the flowering responses of plants was first reported in 1920 by

W. W. Garner and H. A. Allard,[7] who discovered that, after being exposed to appropriate day-length, plant leaves would transmit messages to the buds which caused them to form flowers instead of leafy shoots. They therefore classified plants, on the basis of photoperiodic responses, as 'long-day', 'short-day' or 'indeterminate'. Species of the first two groups exhibit a 'critical day-length'; long-day plants flower when the days are longer than this critical day-length, while short-day plants flower when the days are shorter than this. Indeterminate or day-length neutral plants do not exhibit any critical day-length in their flowering responses.

Photoperiodic induction occurs in the leaves, and the stimulus for flowering moves out to the meristems where flowers are to be initiated. Not only can a partially defoliated plant, such as the cocklebur (*Xanthium pennsylvanicum*), be persuaded to flower by an appropriate photoperiodic induction of the remaining leaf – this is a short-day species – but a single leaf can cause flowering when other leaves of the same plant are maintained under non-inductive conditions. Since sugars, amino acids and other metabolites do not initiate flowering, it is generally assumed that the stimulus must be hormonal. The activity of different wave lengths of light is due to their reversible effects on the pigment, phytochrome, which mediates the many aspects of plant growth and development where red light (640–680 mμ) and far-red light (710–740 mμ) are antagonistic. Thus flowering may be suppressed by a short break of red light in the minimum effective dark period (8.5 hours), but this effect can be nullified by a succeeding exposure to far-red light.[8]

Working with Maryland Mammoth tobacco plants, Garner and Allard were intrigued to find that, at whatever time of year these were sown, they would only flower during the short days of winter. They also discovered that flowering could be induced during the summer by covering the plants for a part of each day, and that winter flowering could be prevented by artificially increasing the daylight hours at that time of year. Later it was shown that the effects of a long dark period could be nullified by a very brief light signal, if it were applied in the middle of the dark period.[9] A short day naturally contains a long dark period but if the latter is interrupted – even for a very brief interval – with light of low intensity, it will be interpreted by the plant as a long day.

At about the same time it was also discovered that the cocklebur could be induced to flower by exposing it to a single short day or, if grown in continuous light, by exposing it to a single long dark period and then returning it to continual light. This suggested that the photoperiodic response might depend upon the length of the dark period. It could be interpreted in terms of an hourglass type of clock in which certain products accumulate in darkness in sufficient amounts to induce a secondary process. Numerous hypotheses have been developed along these lines.[10]

Distribution and photoperiod

The geographical distribution of plants must clearly be affected by the cycles of daylight and darkness as well as by the environmental temperatures that they experience. For example, when tomato plants are grown in a regime of eight hours of light followed by sixteen hours of darkness (abbreviated as LD 8:16), they develop normally; but if the light period is divided into two four-hour periods, each separated by eight hours of darkness, the plants grow very poorly, even though they receive the same amount of light during the twenty-four hours.[11] This can be explained according to the hypothesis of E. Bünning which was outlined in the previous chapter: the second four-hour light period would have fallen in the dark or scotophil stage when light is inhibitory.

One of the most outstanding responses of a plant subjected to an unsuitable lighting regime is that the rate of development of the growing point decreases and may even become disorganized. Mitosis in the apical meristem, or region of active cell division, normally occurs rhythmically, with a maximum just before midnight. No doubt the synchronization of the dividing cells is affected adversely by unnatural experimental photoperiods and, to a lesser extent, by the photoperiods occurring naturally in latitudes to which a plant is not adapted. The endogenous rhythm is not completely independent of temperature – indeed it has a temperature coefficient of 1.25. In most biological processes, the increase in rate produced by raising the temperature 10K (Q_{10}) is between two and three. This may be important to a plant growing at a temperature significantly differing from its optimum and whose internal rhythm requirement is so far removed from the twenty-four-hour daily cycle that it cannot be entrained by the external cycle. This would result in a decrease in growth which could ultimately lead to death.[7]

There is little doubt that plants growing in a particular area are adapted to the local climatic conditions. This is particularly evident in the tropics, where at different altitudes temperatures remain within the same range throughout the year and yet there is a remarkable shift in the composition of the vegetation with altitude. Most plants do not grow over an altitudinal range greater than five hundred or a thousand metres. A series of different plants will therefore be found when ascending from sea level to the snow range in the tropics, as can be done in Peru, Central Africa or New Guinea. This adaptation to local growing conditions is not only a matter of adaptation to temperature, but is also an adaptation to the prevailing twenty-four-hour day–night cycle. For instance the tropical African violet (*Saintpaulia ionantha*) grows optimally on a twenty-four-hour cycle at a constant temperature of somewhat over 20°C. At a higher altitude this plant is not found and presumably could not survive since, at the lower temperature, its optimal cycle length would be considerably more than 24°C. Thus the African violet cannot live at 10°C if it is subjected to a twenty-four-hour cycle of light and darkness, but a thirty-two-hour cycle at that temperature is perfectly satisfactory for normal development. Again, African violets growing normally at lower elevations would die if transported to altitudes of three thousand metres or more; although they might survive at slightly lower altitudes, they would not flower or reproduce.

The distribution of the African violet in the tropics is therefore limited by its response to the light–dark cycle. At higher altitudes, environmental temperatures are so much lower that the endogenous circadian rhythm is slowed down and can no longer be synchronized with the environmental cycle. Consequently, the plant dies within a few months, without ever having been subjected to temperatures near freezing.

The same thing would be true for a plant of higher altitudes which was adapted at the prevalent lower temperatures to the twenty-four-hour cycle. If, for some reason, this plant were to be transported to a lower altitude, its autonomous cycle-length would be shortened so much that it could no longer survive at the higher temperatures prevailing. This has been demonstrated in *Baeria chrysostoma*, whose optimum temperature is about 17°C but which, if kept for sufficiently long at 26°C, actually dies. It has been estimated that about half the response to temperature of tomato plants is due to rhythm desynchronization, and that the death of tropical plants in temperate regions, without them ever nearing freezing point, is also due to desynchronization rather than to biochemical abnormalities.[11]

At the present time there is no evidence that the chemical reactions comprising the biological clock are different in plants and animals.

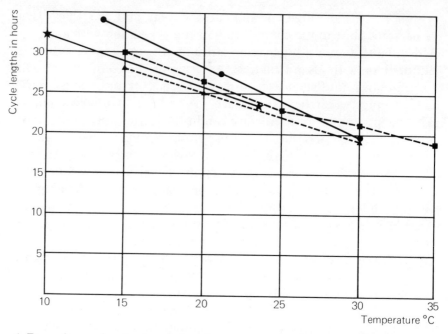

6. Dependence of optimal cycle length on temperature for growth of tomato (triangle), African violet (star) and North American composite (circle), and for nyctinastic movements of scarlet runner-bean (square). (After F. W. Went.)

The observed rhythms, of course, are different but the basic circadian systems are identical, being located at a key level of cell chemistry common to all living organisms.[12] It is, therefore, by no means surprising to find in the animal kingdom situations parallel with those found in plants. For instance, it has been suggested that the geographical range of the yellow fever mosquito (*Aëdes aegypti*) may be limited not, as previously thought, by the range of summer temperature, but by the range of summer day-length which permits a physiological balance between the factors controlling flight activity and oviposition rhythms.[13]

Even should the nature of the biological clock itself be finally resolved, we shall still be far from understanding the full range of the inter-relationship between the clock, the physiological and behavioural systems, and the biotic and physical environment in plants as well as in animals.

Seasonal rhythms in plants

In all but the equatorial regions of the world, climate changes with the seasons. To match these variations plants, like animals, have

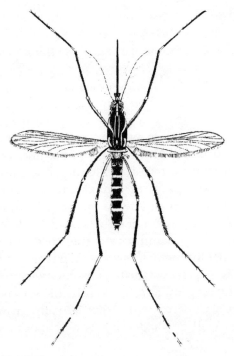

7. The range of the yellow fever mosquito may be limited by the range of summer day-length which permits a physiological balance between the factors controlling flight activity and oviposition rhythms. (From J. L. Cloudsley-Thompson.)

evolved a whole series of adaptive changes which enable them to survive throughout the inclement period of the year. Plants form seeds or tubers, while vertebrates hibernate and insects go into diapause or suspended animation. On the other hand, growth and reproduction take place in the spring when mild weather favours survival of the offspring. A successful plant or animal is one that is already prepared or preadapted for seasonal changes. Thus it cannot be the amelioration of climatic conditions which initiates adaptive changes in the organism. A seed must be ready to germinate immediately the snow melts, or else much of the growing season would be wasted. As we saw in the previous chapter, some organisms adapt to seasonal changes by measuring photoperiod or day-length. Others, in addition, possess circannual clocks which measure time in years.

Beatrice Sweeney tells us that a rigorous demonstration of an annual rhythm has probably never been achieved in plants, for the difficulties of detecting such a rhythm are formidable.[14]

29

Nevertheless, tropical trees sometimes show annual cycles of defoliation even though different trees in the same locality or even different branches of the same tree may lose their leaves at different times.

Many tree species in Central America and other tropical regions have adapted the timing of their flowering and fruiting to coincide with the dry season. In this way maximum use is made of conditions favourable for growth, while pollinating and dispersal agents are exploited to the utmost. No doubt competition occurs between individual trees in mature forest; while the uppermost leaves of the canopy may receive an excess of sunlight, those below certainly do not. Consequently, much of the competition results from lateral growth, since the canopy does not move upward indefinitely, and any factor which reduces the ability of a tree to produce leaves is of importance. The beginning of the dry season is such a factor. During the rains there is a general flush of vegetative growth, and any individual that did not grow as much as possible at that time would be at a disadvantage compared with other actively growing trees; for instance, it might well be shaded by its neighbours, or covered by vines.

Since the dry season is characterized by a lack of vegetative competition, the flowering and fruiting season tends to shift towards that period and reserves stored during the growing season can then be utilized. There is also a tendency for flowers and fruit to be produced as early as possible in the dry season while the soil is still moist. The minimum time necessary for the production and maturation of fruit is also important; maturation before the beginning of the rainy season or at the end of the following rains ensures that seedlings have the maximum chance of receiving sunlight in the early successional stages, and have an entire wet season in which to grow their roots before the next dry season is upon them. In some trees such as the acacias, fruit matures nearly a year after flowering. It may be that the young plants can only store enough reserves for flowering during one wet season, and require another in which to store enough for the maturation of the seeds. Finally, dry air may be of importance in the maturation and ripening of fruit. Many seed pods will only open in dry weather, and the exploding pods of certain plants of the pea family (*Canavalia* spp.) do not open if kept in a humid state.

Pollinating and dispersal agents are also most effective during the dry season. Shortage of water enhances the importance to birds and mammals of fruit and nectar; leaflessness favours the location of flowers and fruit by humming birds, bats and insects. There are substantially more hours of sunshine per day during the dry season – due not to differences in day-length but to the lack of clouds. Many flower-visiting insects, especially bees, show reduced activity during cloudy periods; nocturnal pollinators are likewise inhibited by rain. In the dry season blossoms are not damaged by rain, and the reduced activity of phytophagous (plant-eating) insects enhances the survival of flowers, fruits and seeds. Dry conditions favour ground-nesting bees, for algal growth in nest cells is reduced. In dry weather air temperature rises earlier in the morning than it does during the rainy season, so that there are more hours of the day during which maximum insect activity takes place. Consequently, large complexes of unrelated day-active pollinators build up in the dry season, when there is also a peak in flowering.

Of course not all tropical trees flower and fruit in the dry season; indeed certain other factors tend to promote flowering and fruiting at the time of the rains. No doubt the peak of flowering and fruiting is the result of selection for sexual reproduction at the most opportune time in the year rather than the expression of immutable physiological processes. Plants, like animals, probably respond to proximal environmental factors which synchronize their activities with ultimate ecological considerations.[15]

Adaptive functions

It is clear from the above discussion that seasonal rhythms of growth and reproduction are of adaptive value even to plants living under the equable climate of tropical forest. In temperate and sub-tropical regions, cold winters or seasonal drought impress a periodicity on the vegetation so that winter dormancy or aestivation (the rearrangement of parts on a flower bud) are essential to the survival of most plant species. The advantage of dependence upon a particular day-length for flowering is less immediately obvious than its disadvantages. Throughout the plant kingdom a great variety of dispersion mechanisms may be encountered which are apparently designed to allow different species to move freely across the surface of the earth. Yet day-length requirements may well defeat this end. Perhaps the most distinct advantage of a rather critical day-length requirement is that it ensures that all members of a population will bloom at the same time. This is essential for efficient cross-pollination, and may well arise from selection pressure against flowering at the wrong time. A plant that requires cross-pollination would be doomed if it were to flower by itself.[16] In a comparable way, synchrony of seed production in the piñon pine (*Pinus edulis*) has probably evolved largely in response to predation by invertebrate and vertebrate animals, especially piñon jays (*Gymnorhinus cyanocephalus*) which consume most or all of the seeds from piñon trees which reproduce out of phase with the majority.[17]

Chapter 3

Nature by Night

ONE OF the advantages of being able to measure the passage of time lies in the ability it confers to predict, and therefore be ready to respond to, forthcoming events such as the onset of daybreak or dusk. The transition from daylight to darkness is accompanied by many physical changes in the environment that affect the lives of plants and animals. As night falls there is a drop in temperature, an increase in the relative humidity of the atmosphere and a tremendous decrease in light intensity. Such physical changes are more marked on land than they are in water, and are especially evident in desert regions. In a continental tropical desert, for example, the difference between day and night temperatures may be considerable while in tropical rain-forest it is not more than a few degrees. Mean maximum temperatures in rain-forest seldom exceed 32°C and rarely drop below 21°C, while shade temperatures as high as 56.5°C have been registered in Death Valley, California, and even higher in the Sahara. Only a short distance below the surface of the soil, however, the microclimate becomes progressively more stable. Below about 50 cm. there is scarcely any diurnal temperature variation in the sands of the Sahara. A diurnal temperature range of 30.5K has been recorded on the surface of the soil in southern Tunisia in April, while five centimetres down a cricket hole, the range was reduced to 18K and at a depth of thirty centimetres the temperature range was only 12.5K.[1] When figures such as these are compared with the extremes found on the surface of the desert, the advantages of nocturnal behaviour and of spending the day in a burrow become apparent immediately.

Desert animals

Many small desert rodents, such as kangaroo-rats (*Dipodomys* spp.), can survive indefinitely in their natural habitat on a diet of seeds and

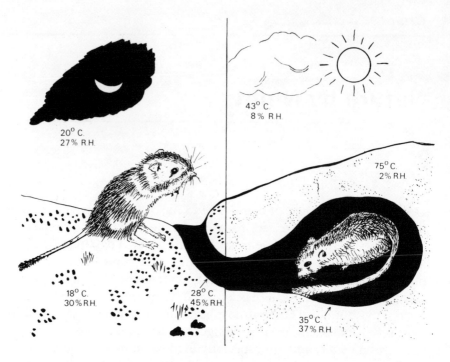

20° C.
27% R.H.

43° C.
8% R.H.

75° C.
2% R.H.

18° C.
30% R.H.

28° C.
45% R.H.

35° C
37% R.H.

8. Typical microclimatic readings above ground and in the burrow of a jerboa at night and during the day. (Drawing: Anne Cloudsley.)

other dry plant materials, for they are nocturnal animals that remain down their holes throughout the day and thus do not need to use water for thermoregulation.[2] Avoidance of extreme temperatures by burrowing and the nocturnal habits so characteristic of desert and savanna animals, is associated with the operation of a circadian biological clock, which tells its owner when the time has come for it to emerge from its hole and begin foraging. But for this, in the comfortable darkness of its burrow, the desert animal might never know when night had fallen on the sands above.

Not only kangaroo-rats, but also jerboas, gerbils, ground-squirrels and many other small animals manage to survive in the desert only as a result of their nocturnal behaviour. Woodlice, centipedes, millipedes, spiders, scorpions, mites, ants, beetles and other insects, as well as snails, lizards, tortoises and snakes avoid the extreme midday heat and drought and the intense ultra-violet light by hiding under rocks and stones or down holes and cracks in the ground, from which they emerge only at dusk. Small birds take refuge in the leaf-bases and young shoots of palm trees, or in bushes and camel-scrub.

Even animals that would naturally be day-active in any other ecological environment restrict their activities to the dusk and dawn, and are crepuscular in the desert environment.[3]

The problems confronting desert species are essentially the same as those encountered in a less acute form by all animals in their evolution from aquatic to terrestrial life. In hot, dry climates, survival depends on avoiding desiccation and keeping cool. Consequently, where water is in short supply there must inevitably be a conflict between the requirements of conserving moisture for vital purposes and of transpiring it for cooling. Small animals can never afford to utilize water for cooling because their large surface-to-volume ratio would result in excessive amounts of moisture being lost, and many of them are able to exist on land only by nocturnal habits and by behaviour patterns that restrict them to microhabitats in which conditions are less extreme than they are in the open. Although larger forms are unable to escape the rigours of the desert climate in the same way as smaller ones, they can better afford to lose water by evaporation because a given rate of evaporation per unit area can proceed for very much longer before the total water content of the body falls to a lethal level. For example, a flea could tolerate a transpiration rate of 5mg. per cm.[2] per hour for about fifteen minutes before losing ten per cent of its water; but a man would have to

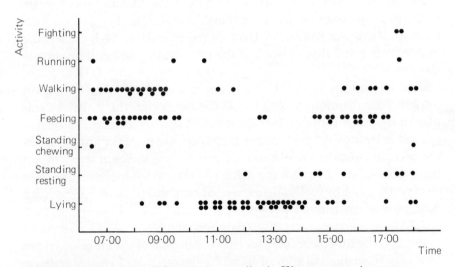

9. Daily cycle of activity of five captive gazelles in Khartoum zoo in summer when midday temperatures reached 45°C or more. The animals were most active between 05.00 and 08.00 hours and again between 16.00 and 18.00 hours. (After L. I. Ghobrial and J. L. Cloudsley-Thompson.)

35

transpire 4,500 times faster for a similar period to suffer the same proportional loss. Nevertheless, even the largest desert animals, such as camels, gazelles, addax and oryx antelope, seek shelter from the blazing sunshine as far as they are able.

Activity rhythms in earthworms

Daily activity rhythms are particularly important in hot deserts, of course, but they are also found among the inhabitants of that most equable environment, the soil. Since the time of Charles Darwin it has been known that earthworms (*Lumbricus terrestris*) have an endogenous rhythm of locomotory activity. Darwin kept earthworms in pots of earth shielded from the light, in his library at Down House, so that he could make periodic observations of their activity. As he often suffered from insomnia Darwin quickly noticed that worms spend the daytime inactive down their burrows underground, emerging at night to feed, mate, and crawl about on the surface of the earth.[4] But he did not comment on the obvious adaptive nature of the cycle of earthworm behaviour which he observed so closely. As Miriam Bennett, who has studied earthworm rhythms for many years, points out, terrestrial worms are constantly threatened by desiccation. Water rapidly evaporates from their moist skins if they are exposed to warm, dry air. Fortunately their timing mechanism tends to keep them in the damp earth during the day and permits them to leave the shelter of their burrows only at night when the humidity is generally higher and the temperature lower than during the day.[5]

The activity rhythm of the earthworm is paralleled by a rhythm of oxygen consumption, and it has been found that the diurnal cycle plays a significant role in the rate at which worms can learn to turn to the left or right when placed in a T-shaped maze.[6] If the animal made the correct choice of turns it moved into damp moss, but an incorrect turn took it into a tube lined with sandpaper where it received an electric shock. The earthworms needed only thirty-two trials to learn the maze when trained between 20.00 and 24.00 hours, but they required no less than forty-five trials between 08.00 and 12.00 hours.[7]

Finally, the time necessary for the withdrawal of the anterior end of the body from an oval spot of light, by dark-adapted earthworms, is less at night (mean 7.13 seconds) than it is at midday (mean 9.93 seconds).[8] This is clearly related to the activity rhythms: the worm's ability to move most speedily is at night when it is likely to be abroad in its natural environment. Whether this is achieved by more efficient functioning of the light receptors at night, by greater muscular efficiency, or through enhanced nervous conduction is not yet known. Since worms crawl faster in the morning and evening than at midday, however, it seems possible that the variations in the time response may depend only upon varying rates of co-ordinated movements.[9]

The possession of a circannual clock would seem to be adapted to the ways of life of earthworms. Certain species, such as *Allolobophora longa* and *Eisenia rosea*, function very obviously in time with the seasons of the year, for they spend the summer aestivating (in a state of torpor) deep in the soil. Garden earthworms, on the other hand, usually mate on damp summer evenings, but their circadian rhythms ensure that they move into the open when their moisture content is likely to be near the maximum.

Nocturnal activity and water conservation

If circadian rhythms play an important role in the ecology of earthworms, they are far more important to many of the inhabitants of more arid habitats. The more primitive terrestrial invertebrates such as worms, woodlice, centipedes, millipedes, springtails and other soil dwellers, avoid desiccation by remaining for most, if not all of the time, in a damp or humid environment to which they are restricted by reflex behaviour mechanisms. These are relaxed only at night when the temperature falls and the relative humidity of the atmosphere increases. Such animals lose water rapidly by transpiration through their integuments because they do not possess a discrete epicuticular wax layer and are consequently only able to venture abroad during the night. In contrast, most of the insects and arachnids are covered by a thin layer of wax which is relatively impervious to water vapour, and thereby reduces transpiration to a minimum.[10]

Of course such a layer is also impervious to oxygen and carbon dioxide. A respiratory mechanism has therefore been evolved which permits gaseous exchange to take place while, at the same time, restricting water loss as much as possible. The spiracles of insects and the lung-books of spiders and scorpions, for example, are normally kept closed by special muscles; only when carbon dioxide in the body begins to accumulate are they opened to facilitate respiration. It can easily be shown experimentally, by weighing the animals before and after subjecting them to desiccation, that the rate of water loss by evaporation is greatly increased when they are exposed to an atmosphere containing five per cent of carbon dioxide. (The amount of this gas present in the air is only about 0.03 per cent.) When larger quantities of carbon dioxide are present, the respiratory apertures remain permanently open and the rate of water loss is considerably increased.

It is probable that the transition of many invertebrate groups from water to land may have taken place via the soil where aerial respiration is not associated with desiccation. Hence the phylo-genetic series – annelids, woodlice and 'myriapods', insects and arachnids – may correspond with the series of environmental changes represented by life in water, soil and air. Possibly the Apterygota, a sub-class of wingless insects, as well as insects and arachnids like scorpions, spiders and ticks, that lack a waxy non-cellular cover and must live in the soil, may represent an intermediate stage in the evolution of terrestrial life.[11] At the same time many immature insects are soil-dwellers and correspondingly nocturnal in habit, with a high rate of transpiration if placed in dry air, whereas the adults are adapted for aerial dispersal and are diurnal in habit. An interesting example of ontogeny apparently recapitulating phylogeny!

Each aspect of adaptation to life on land affects, and is affected by, other aspects. For example, if the integument is rigid and provides support, then growth becomes impossible except by moulting, and this limits size. Size is also limited where the respiratory system consists of tracheae and tracheoles and the physiology of nutrition and excretion are closely concerned with water conservation. Superimposed upon such basic morphological and physiological requirements are the numerous concomitants of behaviour and ecology, for orientation and behaviour mechanisms must also be evolved to retain organisms in environments to which they are suited; to find food, mate and carry out the functions essential to their continued existence, at the most suitable hours of the day or night.

Forms with a high rate of transpiration

Woodlice show a circadian rhythm of activity and tend to wander abroad during the night, when the temperature drops and the humidity of the atmosphere increases. This rhythm is correlated with alternating light and darkness and not with fluctuating temperature and humidity, although these factors of the environment are probably of greater importance in their daily lives. When taken into the open they are stimulated kinetically by light and drought so that they run actively until, by chance, they reach some dark, damp spot in which they come to rest. They also show negative phototaxis, but one searches in vain to find any directed orientation toward moist air or damp surfaces. The animals run aimlessly, turning first one way and then the other. But as the air becomes damper their speed of movement and rate of turning decreases until they finally stop.[12]

By means of choice-chamber apparatus it has been found that the intensity of the humidity response of woodlice is less in darkness than in light and still less in the nocturnal phase, while it increases with desiccation. Movement away from light is more marked in animals which have been in darkness for some time and is still initially shown in dry air, whereas control animals move towards the light when humidity is sufficiently reduced. Finally, the active phase is correlated with increased sensitivity to external conditions.

These experimental results can be related to the nocturnal ecology of the species as follows: woodlice are often to be seen wandering abroad at night when a decrease in the intensity of their responses to humidity enables them to walk in places where they are never to be found during the day. The increased photo-negative response, after they have been conditioned to darkness, ensures that they get under cover promptly at daybreak and thus avoid many potential predators. On the other hand, if their daytime habitat should dry up, the woodlice are not restrained there until they die of desiccation since they tend to become photo-positive in dry air and thus are able to wander in the light until they find some other damp hiding place and again become photo-negative.[13] Even the desert woodlouse (*Hemilepistus reaumuri*) has a circadian rhythm and water-relations essentially similar to those of woodlice from temperate regions. Although this species can withstand hot, dry conditions for some considerable time, there is no 'critical temperature' to indicate the presence of a cuticular layer of water-proofing wax.[14]

Woodlice spend the greater part of their time in an atmosphere that is saturated with water vapour, but there is considerable variation in their ability to withstand dry air and high temperatures, and in the length of time they can remain in dry places. It has been shown that the degree and extent of nocturnalism in various species can be correlated with the ability to withstand water loss by transpiration.[15] Finally, it has been shown that nocturnal emergence is inhibited by wind, because air currents tend to remove the shell of moist air that

39

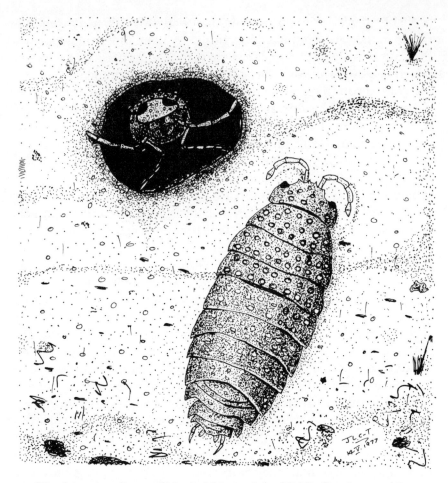

10. The desert woodlouse of North Africa and the Middle East is unusually large with long legs and a curved back. It digs communal burrows in which it remains during the heat of the day. (From J. L. Cloudsley-Thompson.)

surrounds the transpiring animal. The relationship between wind-speed and the number of woodlice wandering abroad at night applies only above a minimum threshold temperature.

Somewhat similar responses have been shown to occur in millipedes, which also lose water rapidly in dry air. Some species may tend initially to move toward dry places, but the reaction is gradually reversed as desiccation proceeds. Orientation is again entirely kinetic and, in an experimental chamber in which a choice of humidities is provided, both the time spent and the distance covered are greater on the moist side. Millipedes possess a clear endogenous twenty-four-hour periodicity and emerge from the comparatively constant conditions within their daytime retreats under stones, bark, logs and fallen leaves, under the stimulus provided by their circadian chronometers. In the case of tropical species, however, locomotory activity is stimulated both by increases and by decreases of temperature, and it is suggested that temperature fluctuations may be of primary importance in the synchronization of the animals' activity with day and night; the effect of light on their activity is slight and this is perhaps an insignificant factor in their natural gloomy habitat in tropical forests.

Like woodlice, millipedes avoid the light but with the exception of a tactic response in those forms that possess eyes, their reactions are non-directional. When illuminated they crawl around until they find themselves in darkness, where they come to rest. Although to millipedes, as to woodlice, humidity is the most important factor of the environment, these animals are not able to find the way directly to damp places, instead they are merely repelled kinetically by drought. Nevertheless this stereotyped and curiously negative behaviour is surprisingly effective in preventing them from wandering away from their normal habitats; but it does raise the problem of how dispersal can take place and new habitats become colonized. There are a number of cases on record of millipedes, sometimes accompanied by centipedes and woodlice, migrating in vast armies. Occasionally they have crossed railways and been squashed in such numbers that locomotives have been impeded and sand has had to be strewn on the lines before their driving wheels would grip. At other times cattle have refused to graze on invaded pastures, wells have been filled with drowned corpses, and workmen cultivating the fields have become nauseated and dizzy from the odour of millipedes crushed by their hoes. Such mass migration, however, is of rare occurrence and local in extent, so that the net effect on distribution is probably negligible. The explanation of the problem of distribution lies partly in the fact that the restraining mechanisms are somewhat relaxed during the night when the temperature falls and the humidity of the air increases. Thus it is at night that these creatures are enabled to disperse themselves and to overcome the restrictions inherent in the physiology of the integuments.

Another example of the way in which physical conditions may play an important part in the timing of circadian rhythms is afforded by the adaptive divergence in water-relations of fruit-flies. These are closely correlated with the daily cycle of atmospheric moisture conditions, despite the fact that the insects are genuinely terrestrial and presumably have moderately efficient epicuticular waxes. At the time of eclosion, however, fruit-flies are very susceptible to desiccation, and their wings commonly fail to expand in low humidities. Eclosion is correspondingly restricted to the cool and moist time near dawn. Although this behaviour pattern is adapted ultimately to moisture conditions at daybreak, it is controlled proximally by entirely different variables: an endogenous clock synchronized by the dark-light transition at dawn. Furthermore the flying, feeding and sexual activity of the mature flies are virtually restricted to the hours from dawn to sunset. Yet even between closely related species there may be variation in the actual time of emergence from the puparium, which depends upon relative transpiration rates.[16]

In England, the fruit-fly *Drosophila subobscura* has a marked peak of activity in the evening which is highly correlated with light intensity, while *D. obscura* shows gradually increasing activity during the day. Thus it seems that *D. obscura* responds to low light intensity and to low temperature, but has a wide range of tolerance to both light and temperature, while *D. subobscura* responds to low and possibly high temperatures, but primarily has a very definite light optimum.[17]

At first sight it may appear curious that the Drosophilidae should be crepuscular in habit, because other flies of the sub-order Schizophora are mainly day-flying insects. The Drosophilidae have apposition eyes which are usually associated with good directional vision in daylight, and *D. subobscura*, in particular, cannot mate without light, at least in the laboratory. However, in the wild type *D. melanogaster* which have a similar periodicity there is a shift in the spectral sensitivity of the eye to shorter wave-lengths when light intensity is low, and this tends to increase efficiency at dawn and dusk. Species of Drosophilidae that inhabit humid, sub-tropical forests have a more flexible rhythm than those found typically in open country. For this reason it has been suggested that *Drosophila* spp. may have become adapted to increasing daytime aridity, not by increased resistance to desiccation as must have happened in most other Schizophora, but by increased visual efficiency at low light intensities together with a shift in flight periodicity and eclosion to climatically less extreme but darker times of day.[18]

A relationship has been established in the dusky cockroach (*Ectobius lapponicus*) between the amount of activity and the relative humidity of the atmosphere. The males are active during the afternoon, the females and nymphs after sunset. Activity is greater in the evening and increases with higher humidities.[19]

Studies on the ecoclimatic divergence of the mosquitoes *Anopheles bellator* and *A. homunculus* in Trinidad rain-forest provide another example of an inter-relationship between climatic factors and insect periodicity. In both species the shape of the curve of the biting-cycle throughout the hours of daylight parallels closely the curve for atmospheric humidity. Air is moister near ground level, but the humidity gradient varies during the day so that the mosquitoes move up or down and remain in a fairly constant micro-environment. *A. homunculus*, however, requires a higher humidity than *A. bellator*; it always remains nearer the ground and its morning drop of activity precedes that of *A. bellator*, while its evening rise begins later.[20]

The majority of other terrestrial invertebrates are not well adapted to life on land and can be likened to woodlice, centipedes and millipedes in that they can survive only in comparatively moist surroundings. Examples are afforded by triclads, nematodes and earthworms. Molluscs do not appear to have been much investigated from this point of view, but clearly a land snail withdrawn into its shell or a contracted slug can be likened to an arthropod with a wax-layer, while its water-relations, when it is crawling about, will resemble those of a woodlouse. Consequently terrestrial molluscs tend, like woodlice, to be active mostly at night and after rain. As in arthropods, locomotory activity in slugs is stimulated by falling temperatures, but once they have become adapted to light their activity is not affected by illumination.

Like the animals discussed above, amphibians, with their moist skins, are extremely vulnerable to desiccation. Recent work on the common African toad (*Bufo regularis*) has demonstrated the existence of an endogenous circadian rhythm which, although not particularly clearly marked, nevertheless persists for some days in constant laboratory conditions. In an environment of daylight and darkness, with normally fluctuating temperature, toads are almost entirely nocturnal in habit. This is probably an adaptation, not only to the time of activity of the insects on which the toads feed, but also to their water-relations. The African toad has not become adapted to tropical conditions in the ability to conserve water by a reduction in transpiration, as have certain other species, but in Egypt and the Sudan is restricted to the Nile valley and to places where there is perennial moisture. Even here evaporation is extremely high during daytime throughout most of the year, and toads would almost certainly not survive unless their activities were confined to the hours of darkness.[21] Except during the wet season, African toads are only seen at night in Tanzania; the days are spent in secluded spots protected from the extreme hazards of the climate, as well as from possible predators, but at night they become active and migrate from pool to pool.[22]

The above remarks should not be interpreted as implying that in cooler regions of the globe amphibians are not active during the hours of daylight. Indeed sun-basking has been noted in all classes of terrestrial vertebrates, including amphibians, and it is clear that behavioural thermoregulation does occur to some extent in these animals. Nevertheless, although the behaviour of toads may be effective in elevating their body temperature above that of the ambient air during the day, such thermoregulation, limited as it is by excessive evaporation, is not comparable with that found in arthropods and reptiles.

Forms with a relatively impervious integument

The conflict between the incompatible requirements of respiratory exchange and prevention of water-loss in arthropods possessing a discrete cuticular wax-layer has been illustrated by a comparison of the mesh-webbed spiders *Amaurobius (= Ciniflo) ferox* and *A. similis*. Although there is some overlap in their territories, *A. similis* tends to inhabit somewhat drier environments than *A. ferox*. Experiments with aktograph apparatus, by which their activity is recorded, indicate that both species are nocturnal in habit, over ninety per cent of their activity taking place during the hours of darkness. When their water-relations are considered, however, it is found that there is a critical temperature at approximately 35°C above which both species quickly lose water by evaporation in dry air, but below this *A. ferox* loses water more rapidly through its lung-books than does *A. similis*. The rate is almost doubled in both species when ten per cent carbon dioxide is present, as this keeps the lung-books open. Conversely, the length of time of survival in air of fifty per cent relative humidity, and in dry air, is longer in *A. similis* than in *A. ferox*, death ensuing when from twenty to twenty-five per cent of total weight has been lost by evaporation.

A. similis tires more rapidly than *A. ferox* when forced to run at full speed without stopping, but both species can run for long periods when supplied with oxygen. *A. ferox* becomes anaesthetized more quickly in ether vapour and has the larger number of leaves in its lung-books. Its greater stamina, therefore, depends upon a proportionately larger respiratory surface acquired at the expense of great dependence upon environmental humidity.[23]

When arthropods possessing an impervious integument are considered, the adaptive significance of nocturnal habits is almost as difficult to understand as it is in the case of larger animals such as vertebrates. In some instances the physical factors of the environment, especially temperature and humidity, play a dominating role. In others biotic factors seem to be of greater importance, and these will be discussed in subsequent chapters. Here we are concerned mainly with activity rhythms that are adapted primarily to nycthemeral physical changes in the environment.

Behavioural thermoregulation

Most animals show some degree of thermal adjustment, but only some of the arthropods and reptiles, whose integuments are comparatively impervious, are able to maintain body temperatures consistently higher than the temperature of the environment. Evaporation from the moist body surfaces of the other terrestrial poikilotherms (cold-blooded animals) precludes the attainment of temperatures comparable with those of homeothermic mammals and birds. Not only are high body temperatures unattainable, but high temperatures generally, insofar as they increase the saturation deficiency, or evaporative power, of the air may be positively harmful. Nevertheless, probably all animals that are capable of movement to some extent regulate their body temperatures by avoiding excesses of heat or cold.

On account of their very small size, metabolic heat production can be effective in Arthropoda only for limited periods. Some insects show warming-up movements and flutter their wings before flight. It has been found, for example, that the sphinx moth maintains a mean temperature of 38° when the temperature of the ambient air is 16°C. The insect does not take off voluntarily when its body temperature is less than 32°C, although it is capable of strong flight at thoracic temperatures as low as 25°C. The ability to begin and cease flight fairly abruptly results in an economy of energy and compensates in part for the small size and unfavourable surface-to-volume ratio of the moth. In this way, far less energy is expended than would be necessary to maintain a constant high thoracic temperature. Yet in flight the thermal efficiency of the moth is comparable with that of a bird.

Metabolic heat is a much more important factor in temperature regulation among social than among solitary insects, on account of the cumulative effect of the large numbers of individuals present. It has been found, however, that even individual bees show increased metabolism at temperatures below that of the hive. The insects also exhibit a diurnal rhythm; at night they are truly poikilothermic and show no compensation for changes in the temperature of the environment.[24]

Arthropods and reptiles are ectothermal. They achieve thermal control by behavioural rather than by physiological mechanisms which are extravagant when the surface to volume ratio is high. Since they do not transpire rapidly they are not restricted to nocturnal activity. On the contrary, by moving into the sun or shade; orientating the axis of the body to the angle of the sun or the direction of the prevailing wind; increasing or decreasing contact with the soil; perching on branches or entering burrows, day-active reptiles, and to a lesser extent insects, are able to achieve varying degrees of thermal homeostasis.

Thermoregulation by sun-basking and the avoidance of excessive heat, supplemented by more subtle parameters of behaviour, is intimately associated with rhythmical activity. An endogenous periodicity is responsible for emergence at dawn or dusk, independently of cyclical environmental changes in temperature or light intensity. But for this, ectothermic arthropods and reptiles would be unable to make full use of their burrows and other retreats without losing valuable time at the commencement of their normal period of activity. A degree of homeostasis is thus achieved by the interaction of external factors, cyclic processes and behavioural responses.[25] The Nile crocodile (*Crocodylus niloticus*), for example, maintains a relatively constant body temperature by spending the night in water, basking in the morning and evening, and retreating into the water or shade during the heat of the day.[26]

Whereas the endotherms appear to regulate their body temperatures around a single set point, ectotherms have both high and low points, separated by several degrees centigrade. They commonly undergo fluctuations in body temperature, even though these are less than occur in the ambient environment, and cannot rely upon a rigid internal receptor or receptors. In general preferred temperatures tend to lie in the upper regions of the normal range of activity. This observation can be explained by the fact that metabolic efficiency increases at higher temperatures up to the point at which enzymic and other metabolic processes begin to suffer damage or disorganization from excessive heat.

Behavioural control of body temperature implies the existence of an absolute sense of temperature. Virtually nothing is known of the mechanism on which this sense depends, although it apparently lies in the anterior hypothalamus of the pituitary of reptiles. (No comparable thermosensitive region has been described in arthropods.) It probably evolved among the earliest dry-skinned vertebrates during Permian times and before the emergence of warm-blooded homeotherms. It is acted upon both by afferent impulses from temperature receptors at the periphery of the body and by small changes in the temperature of the blood. By this means, the regulation of body temperature is correlated with blood-pressure, respiration, endogenous rhythms of activity, the sexual cycle and the control of metabolic and endocrine functions.

From the above discussion it will be seen that the physiology of circadian rhythms is closely interrelated with temperature regulation and associated functions such as blood shunting from superficial to deep tissues as a means of heat conservation. Circadian rhythmicity is, therefore, one of the many inter-related functions by which an animal becomes adapted to its environment.

Sleep

The physiological necessity of sleep to animals is familiar, if unexplained.[27] It is natural, however, that the alternation of periods of activity with repose should be correlated with day and night.

Sleep is a true instinct in mammals, controlled by the hypothalamus which is the brain centre responsible for instinctive behaviour. Among insects and other invertebrates akinesis or immobility may or may not be associated with deep sleep.[28] It is possible to distinguish between akinesis and sleep only by studying their respective intensities and rates of metabolism.

Little is known of the sleeping habits of wild animals. Many species sleep in characteristic poses which we seldom see because the creatures are easily disturbed and react by flight. For example, crayfish and snakes assume rigid and bizarre postures.

Fishes often sleep on the bottom of a pond or stream; birds tuck their heads under their wings; and bats hang upside down. Whales, sea-lions and some seals sleep under water, only coming to the surface to breathe. Many small mammals such as the dormouse, potto and some sloths sleep curled up in a ball, while the fox uses his bushy tail as a pillow. Elephants sleep only for a couple of hours at about midnight; a healthy elephant may lie with its trunk coiled up like a rope and its head resting on a pillow of vegetation. Elephants that are sick or upset, however, do not lie down at all, but merely doze for short periods whilst standing up. Giraffe, too, sleep only for a few hours holding their heads erect; occasionally and for brief intervals they may rest them on the ground or on their backs.[29]

The need to sleep is probably not only due to fatigue, but also reflects the activity of an animal's biological clock. It is a physiological solution to the problem of how to keep still and quiet in a safe place, without getting bored; dreaming meanwhile may prevent the delicate mechanisms of the brain from deteriorating through nightly disuse. During sleep urine flow decreases and body temperature falls, while hunger and thirst are diminished. These functions are controlled by the circadian clock so that they do not interfere with sleep.

The most striking manifestation of sleep is the partial or complete depression of the higher nervous centres, a state which is subjectively described as unconsciousness. It is only animals with a comparatively high level of mental development which show sleep comparable to that of man, for it is only the higher parts of the central nervous system that are involved. As a result of muscular relaxation metabolism is reduced, the temperature drops and respiration becomes slower and deeper. The rate of heart-beat decreases and the concentration of carbon dioxide in the blood becomes greater. At the same time digestive processes continue in a normal manner and all kinds of stimuli are capable of producing physiological effects on the sleeper without waking him.

It has been suggested that sleep represents a state of inhibition always present in some areas of the brain, but which has now become diffused throughout the entire cortex, even descending to some of the lower parts. The details of experimental results are in agreement with this interpretation. For example, prolonged stimulation of the same region of the cortex results in mental depression and lethargy, whilst a great number of quickly changing stimuli tend to keep an animal alert. However, there is no sharp dividing line between sleep and wakefulness and the state of alertness is probably only an expression of a preponderance of excitation. It is probable that in mammals a rise of cortisol in the morning causes wakefulness, while a fall in the evening brings about sleep.[30]

Since Lucretius wrote *De Rerum Naturae* it has been believed that sleeping dogs may dream. A dog will make chewing movements with its jaws if food is placed before it whilst it is sleeping. A veteran horse which served in the Turko-Italian war, in later life would neigh excitedly in its sleep and kick with its hooves as though living again the scenes of battle. Sleeping chimpanzees sometimes indulge in wild howling, and it has been suggested that this may be caused by nightmares. But the proof of subjective phenomena such as dreams in animals may remain for ever beyond our grasp.

It is said that by dreaming, man and possibly other mammals are able to resolve the psychological conflicts that confront them in everyday life. Dreams are therefore greatly beneficial, and human subjects who are deprived of them under experimental conditions tend to become neurotic. Thus even this aspect of sleep has an important adaptive function, and helps in adjustment to life in an environment where time passes, not evenly but in rhythmic cycles.

Chapter 4

Preadaptations and the Interaction of Environmental Factors

HOMEOSTASIS, THE maintenance of constancy in the internal environment of living organisms, represents one of the major evolutionary achievements of adaptive physiology. However, as we now understand it, homeostasis implies more than this. By means of its circadian rhythms the living organism mirrors internally, or even preadapts to, diurnal changes in the external environment. Claude Bernard's concept of homeostasis was formulated in 1878, and to this day remains a cornerstone in the understanding of physiology. It should nevertheless be modified; homeostasis does not necessarily imply a lack of change because the steady-state, to which the regulatory mechanisms are directed, shifts with time.[1]

This notion can be applied to organizations at cellular, organ-system, individual and social levels. It may also be considered in relation to time intervals ranging in duration from milliseconds to millions of years, although in the present chapter only its significance over periods of twenty-four hours will be discussed.

The circadian rhythm is constantly modified by seasonal changes which may also be mirrored internally and will, at the same time, be influenced by environmental changes in the photoperiod, ambient temperature, and so on. For example, day-active beetles and lizards often show two peaks of activity each day in summer, when the weather is very hot at noon, but only one during the spring and autumn months, while the winter is spent in hibernation. Such seasonal changes in the circadian rhythm may persist under constant conditions. Seasonal changes may also occur in their 'eccritic' or preferred temperatures.

Chronobiology

Because biological systems change rhythmically, it follows that living organisms must be biochemically different during different phases of

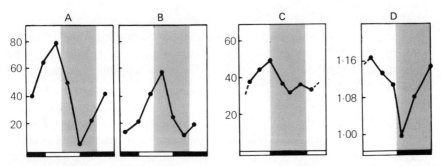

11. Circadian susceptibility of mice to drugs. Ordinates: percentage mortality when injected intraperitoneally with A, *E. coli* endotoxin and B, ethenol (after F. Halberg *et al*); C, Survival time (hours) of mice injected intraperitoneally with librium (5.4 mg. per 20 g.b.w.) (after E. Marte); D, Relative potency of pentobarbital injection with sleep duration as a measure of response (after F. Halberg). Abscissa: time of day. Shaded areas: darkness.

the day and night. Consequently their reaction to the same stimulus will vary at different times. This differential response to an identical stimulus has been confirmed repeatedly with respect to a variety of stimuli including drugs; poisons and other chemical substances; physical agents such as noise and radiation; and biological agents such as endotoxins.[2, 3] Circadian variations in the response to such stimuli may be quite dramatic. For instance, the duration of sleep resulting from an identical dose of pentobarbital sodium averaged 104 minutes when administered at one phase of a rat's circadian rhythm, but only 43 minutes at another. Cystosine arabinoside is also far more toxic at one phase of the circadian system of the mouse than it is at others, while a fixed dose of amphetamine which killed 93.4 per cent of experimental animals when administered at one phase was lethal to only 22.4 per cent at another.[4] When mice were maintained under an artificial-light regime of twelve hours of light followed by twelve hours of darkness, no less than eighty per cent were killed by a standard dose of *Escherichia coli* toxin, injected towards the end of the light period. Under twenty per cent succumbed, however, to a similar dose administered in the middle of the dark period.[5]

Similar results have been obtained with intraperitoneal injections of ethanol, and a comparable rhythm in the human metabolism of alcohol has also been demonstrated[6] – a fact that many people will have observed subjectively! Seizures in mice, produced by the sound of a bell tube, are most common at the time of maximal locomotory activity, which occurs shortly after the onset of darkness when the glycogen content of the liver reaches its lowest level. To the extent that glycogen can be taken as a resource for energy expenditure, even this rhythm may be important for survival,[1] as is the timing of birth, one of the most critical events in the life of an individual. The functional significance of daily rhythms in the resistance of amphibians to heat is unknown; it may be the product of some underlying physiological rhythm, or even have survival value in its own right. A daily rhythm of heat resistance in frogs (*Litoria caerulea*), for example, could be of importance during the summer, after periods of heavy rain, when the animals are often found in temporary pools.[7]

A diurnal rhythm of sensitivity to toxic or other harmful environmental variables is not necessarily advantageous but, when the strength of such variables alters rhythmically, the diurnal rhythm can be seen to be an adaptive specialization. Circadian periodicity, as an adaptation to the environment, involves biotic as well as abiotic factors. Timing of activity, with regard to microclimatic conditions, can be as important as is the correct phase relationship between predator and prey, or the synchronization of sexual partners – topics that will be discussed in subsequent chapters.

Sensory adaptation

In many species, there is a marked specialization for a nocturnal or diurnal habit. Diurnal animals have become adapted to meet, or to avoid, relatively high temperatures and evaporation rates, bright light and decreased conductivity of the air for odours. Conversely, nocturnal species are adapted to decreased temperatures, high humidity, dim light, and increased conductivity of the air for odours. It has been suggested that some crepuscular species may be confined to the short periods of dusk or dawn, simply because they are relatively unspecialized in these respects.

In strictly nocturnal or diurnal species there is usually a refinement of one or more of the senses and differences are of degree rather than of kind. Scent is perhaps most important at night for the congregation of individuals, sex attraction, following the tracks of prey, location of enemies, and so on; hearing and sound production for communication, the detection of enemies or victims; and light production for sex attraction and warning. Diurnal periodicity in sound production at dusk and dawn has been studied in tropical rain-forest fauna. There was found to be a correlation between sound production by diurnal and nocturnal animals (cicadas, orthopterons, frogs, birds, and monkeys) and the physical influences, such as daylight, operating on them.

The sensitive elements in the retina of the vertebrate eye are of two kinds – rods and cones. Animals that are adapted for night vision have a particularly large number of rods. In man, the central part of the retina, the fovea centralis or yellow spot on which objects are normally focused, consists almost entirely of cones. These are responsible for colour vision and for detailed resolution. Towards the periphery of the retina the proportion of rods increases; that is why a very faint star can often be seen only when one looks slightly to one side of it, a principle frequently employed by astronomers. Colour-blindness in nocturnal animals such as mice is associated with lack of cones.

In the eyes of nocturnal animals a reflecting layer, called the tapetum, is usually present. Any light passing through the retina is reflected back so that there is a double chance for a given ray to stimulate a rod. The absence of a tapetum in owls illustrates the super-efficiency of the retinal rods through which little or no light passes. There are many kinds of tapetum, but reflection is most frequently caused by fibrous tissue or guanin crystals. Associated with a very efficient retina, as in the opossum, a tapetum results in such sensitivity that bright light is avoided at all times, even though a slit pupil very greatly reduces the amount of light penetrating to the retina, and ensures some protection by means of its fine control.

The larger terrestrial mammals – ungulates, elephants, lions, bears, and so on – have eyes which are not particularly specialized either for day or night vision. They have a tapetum and sufficient rods for reasonable visual sensitivity, but they also have many cones. These, coupled with a large eyeball, ensure good resolution of detail. In consequence, vision is relatively good both at night and by day, even where scent and hearing are also acute, and the animals are not very markedly nocturnal or diurnal.

Nocturnal forms often have an iris with a vertical pupil aperture, and adaptations may also occur in the lens. For example, flying squirrels are nocturnal and have colourless lenses, whereas true squirrels (*Sciurus* spp.) are diurnal and have yellow lenses. Both types inhabit the same forest in many localities, and their different periods of activity are related to the light-filtering properties of their eyes.

In diurnal eyes intra-ocular oil droplets act in the same way as filters in photography. The red droplets in birds are effective at sunrise but, as the day wears on, the yellow and colourless droplets are most effective; at sunset the red droplets again come into operation. Thus late-rising birds such as hawks have fewer red droplets than song-birds. Red droplets enable turtles to see through the surface glare of a tropical sea.

Yellow filters cut out much of the violet light, and some blue. These are the colours responsible for most chromatic aberration. At the same time they enhance contrast and do not impede natural hues. The combination of a yellow diurnal lens or yellow droplets with a tapetum would result in poor vision at all times, and does not occur in nature.[8]

The majority of spiders have six or eight eyes, which appear to be of two kinds; some are black, others pearly white. They were at one time described as diurnal and nocturnal respectively, and it was believed that one type was used in daytime, the other at night. The evidence for this is slight. The chief visual organs of insects and crustaceans are the compound eyes, made up of a number of transparent facets each with an elongated, light-sensitive structure beneath it. In diurnal insects images are formed by apposition – that is, the light passing through each facet falls only on the appropriate optic rod or rhabdom. In many nocturnal insects such as noctuid moths, glow-worm beetles and the like, there is a special kind of compound eye that forms images by superposition. Each sensory rhabdom receives light rays not only through its own facet, but through neighbouring facets also. Between these two extremes there are many intermediate stages. Light and dark adaptation of compound eyes is achieved by migration of the pigment in each optical unit or ommatidium, so that the rhabdom is more or less shielded. Similarly migration of rods, cones and retinal pigment is of great importance in the lower vertebrates.

There is an Arabian fable concerning an argument between the horse and the lion as to which had the keener sight. The lion could distinguish a white pearl in milk on a dark night, but the horse could see a black pearl in a heap of coal in the day. The matter was submitted to arbitration, and the judges rightly pronounced in favour of the horse. No doubt they realized that the acuity of the cone vision of ungulates is greater than that of the nocturnal rods and tapetum of carnivores. Old-time Arab horse-breeders are said to have invented a competition to establish whose horse could see the greatest distance. While the masters waited together in a great semi-circle, the horses were released from a considerable way off, upwind. The winner was the horse who first recognized its master and galloped back to where he stood.

Parasite rhythms

Since the time of Hippocrates it has been known that patients suffering from malaria experience rhythmic bouts of fever, while the periodicity of microfilaria has been known for nearly a century. More recently other parasitic rhythms have been recognized and their functions elucidated. According to Frank Hawking, the circadian rhythms of parasites may be grouped as follows:

(a) Rhythms that depend on the synchronous cell division of the parasites (e.g. malarial parasites).
(b) Synchronous ejection of infective forms from the body of the host at some particular time of the day or night (e.g. ova of schistosomes).
(c) Rhythms in which the same individuals migrate backwards and forwards in the body of the host (e.g. microfilariae).
(d) Migration of worms up and down the intestine on a twenty-four-hour pattern (e.g. cestode tape-worms).

In each case the expression of these rhythms results in the pre-adaptation of the parasite so that it may take full advantage of some forthcoming event.[9, 10]

(a) *Malarial parasites* The mature male gametocyte of the malarial parasite is a short-lived organism. If it is not taken up by a mosquito within a period of not more than about six hours, it breaks down and has to be replaced. The same is probably true of the mature female gametocyte. There must therefore be strong selection favouring the concentration of the brief period of maturity into the hours when the mosquitoes which transmit the parasites normally take their blood meals. The asexual cycle of the life history is consequently adjusted so that sexual forms appear at a favourable time. The rhythm of the parasites is apparently entrained to the nycthemeral periodicity of the host, and the synchronizer seems to be the host's cycle of body temperature.[11] There is probably a feedback connection, however, between parasitic infection and body temperature. Thus the attack of fever which follows schizogany alters the timing of the normal cycle of body temperature and greatly increases its range. This would explain why the cycle of the malarial fever induced by infections of *Plasmodium vivax* is often shorter than forty-eight hours, so that attacks tend to occur earlier and earlier in the day.

(b) *Schistosomes* The infective forms of certain parasites (such as oocysts of the coccidea of birds and the ova of *Schistosoma haematobium*, which causes bilharzia in man) are discharged

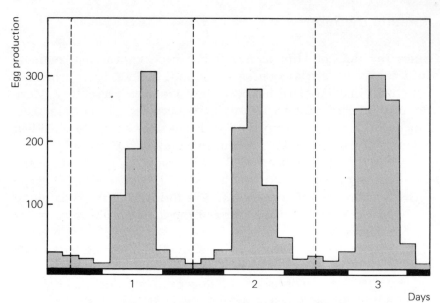

12. Output of eggs of *Schistosoma haematobium* in man. Ordinate: egg production expressed as percentages of the mean for the whole three days. Abscissa: time. The black strips represent the period from 18.00 to 06.00 hours. (After F. Hawking.)

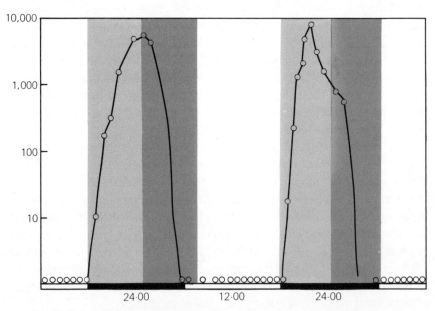

13. Excretion of oocysts of coccidia from a protozoan parasite (*Isospora larazei*) from a sparrow over a period of two days. Ordinate: number of oocysts per gram of faeces (logarithmic scale). Abscissa: time. The cycle of illumination consisted of daylight shining into a room (07.00–18.30 hours); weak lamplight (18.30–24.00 hours); darkness (24.00–07.00 hours). (After F. Hawking.)

rhythmically. In neither of these cases has the timing mechanism yet been elucidated but it is possible to speculate on its functions. The oocysts of the coccidia of sparrows, for instance, are discharged almost exclusively in the late afternoon and early evening, when the birds bathe in the dust and preen their feathers before roosting. At this time, the oocysts presumably contaminate their beaks and are accidentally ingested. If they were discharged during the day, when sparrows disperse widely for feeding, they would have little opportunity of being ingested by another host.

In a similar way, the number of schistosome eggs excreted in the urine of bilharzia sufferers shows a distinct peak in the morning. If the ova fall into fresh water they hatch out into miracidium larvae which must burrow into the bodies of aquatic snails if they are to continue their development. No doubt the molluscan hosts are more readily available in the light and warmth of the day than they would be during the cold dark night.

(c) *Microfilariae*   These are the larvae of filarial worms – long, thin nematodes which become embedded in the tissues of man and other mammals, birds and poikilothermic vertebrates. Many species show diurnal periodicity, and the microfilariae are numerous in the blood of the host at certain times of day or night but scarce at others. For instance, the microfilariae of *Wuchereria bancrofti*, the organism that causes the dread disease of elephantiasis in man, are usually very numerous in the blood at night but rare or absent during the day. *W. bancrofti* is transmitted by mosquitoes which bite at night; microfilariae of the Pacific type of *W. bancrofti*, however, though present in the blood at all times, are especially numerous in the afternoon. This parasite is transmitted mostly by day-biting mosquitoes. In the case of *Loa 'loa*, the filarial worm that causes calabar swellings, microfilariae are most numerous in the blood during the day. They are transmitted by tabanid flies (*Chrysops* spp.) which bite in daytime. Finally, microfilariae, which are most numerous in the peripheral blood-stream in the evening or at night and are less numerous in the early morning (e.g. *Dirofilaria immitis*, a parasite of dogs), are transmitted by mosquitoes which bite during the evening or at night.

In all these examples, the microfilariae are most numerous at the time when the insect vector of the disease sucks blood, and this part of the cycle is clearly a preadaptation that enhances the parasite's chances of transmission. At times when microfilariae are absent from the peripheral blood, they are to be found in great numbers in the

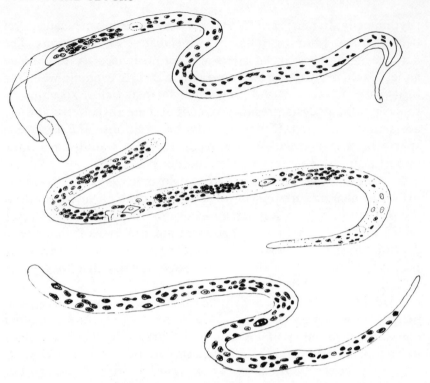

14. Microfilariae, the tiny worms which cause elephantiasis, calabar swellings, onchocerciasis and other dread tropical diseases show rhythms of activity and appear in the peripheral blood streams of the host at the time when the blood-sucking insect vector is likely to be seeking food.

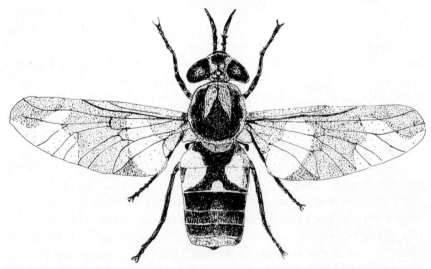

15. A day-active horse-fly (*Chrysops* sp.) which transmits the filarial worms that cause calabar swellings in man.

small vessels of the lungs.[12] Here they accumulate as a result of a positive response to increased oxygen tension. Dispersal throughout the blood occurs during the passive phase of the circadian rhythm. It is suggested that the periodicity of the microfilariae represents a compromise between the advantages of occupying a position in the lungs, and the necessity of transmission to new hosts. Experiments on the reactions of microfilariae suggest that, during their active phase, they respond to increased oxygen tension by reversing their movements so that they resist the blood flow, and accumulate in the pulmonary arterioles. When they are passive, however, they are carried through the lungs and become dispersed throughout the blood. (In the Pacific type of *W. bancrofti* an opposite pattern of behaviour must operate.)

(d) *Tape-worms* Evidence has recently been obtained showing that certain intestinal worms migrate forwards during part of the day or night and then slip backwards again during the remainder of the twenty-four hours. In particular, the tape-worm *Hymenolepis diminuta*, a parasite of rats, migrates forward during the night when its host is actively feeding and moves back during the day when the rat is normally passive.[13] Presumably the significance of such migrations is related to the nutrition of the parasite.[14] This would appear to be the first report of a circadian rhythm, in a parasite, that is not associated with the facilitation of transmission to a new host.[9] No doubt it is an adaptation to the availability of food; transfer to a new host is an essential factor in the life-cycle of any parasite, but nutrition is scarcely less important.

Insect emergence rhythms

Considerable survival value stems from the operation of biological clocks in timing the physiological activities of living organisms. Examples were given in the previous chapter of locomotory rhythms that are correlated with the avoidance of high temperatures and low humidity during the hours of daylight. Another way in which endogenous rhythms can be used to protect animals from desiccation is seen in the emergence of adult fruit-flies and other insects from their puparia. Although the formation of puparia in cultures does not take place rhythmically throughout the day and night, eclosion tends always to occur shortly after dawn. In the case of *Drosophila pseudoobscura*, populations show daily peaks of emergence from one to two hours after dawn, whereas in *D. persimilis* the peak occurs about four hours after daybreak.[15]

Culture of *Drosophila melanogaster* display no emergence rhythm when they are maintained in continuous darkness but, if exposed to a brief period of light at any time during larval or pupal development, subsequent adult emergence will occur synchronously. This phenomenon has been investigated in depth, but mainly in order to elucidate the nature of the biological clock mechanism.[16] Light and dark stimuli have different timing effects on emergence. Both of these phase-setting synchronizers normally occur daily during the nycthemeral cycle, and the time of day at which adult emergence occurs in nature depends on day-length, the result of the temporal relationships between dawn and dusk.

The adaptive function of swarming after emergence may be related to the suitability of environmental factors such as temperature, humidity or light intensity. Swarming may also be related to sex and other aspects of behaviour.[17] The interactions of the various factors which affect emergence rhythms are extremely complex. They include the critical day-length for the induction of diapause, the developmental stage at which diapause is determined, the growth stage at which the response is manifested, the effect of temperature on photoperiod, and the general thermal characteristics of growth.[16]

Interaction of environmental factors

Under natural conditions, the activities of animals are influenced by many environmental factors in addition to photoperiod. These include temperature, humidity and light intensity, as well as physiological factors such as age, and stage in the reproductive cycle.[16] Stress may also lead to a modification or reversal of activity patterns. For example, the ground beetle *Feronia madida* is normally nocturnal, but may adopt day-active habits under conditions of starvation. In woodland where there is a rich nocturnal fauna its rhythm of activity is reinforced by the successful capture of prey so that it remains strictly nocturnal. In open country, on the other hand, where nocturnal animals are less common the ground beetle is not inhibited by light when hungry and becomes diurnally active for as long as it is successful in catching food during the hours of daylight.[18]

Again, although most of the movements of harvestmen (Opiliones) occur at night, when they are hungry, they may also be active during the day, especially if the humidity is high. As a result of his investigations into the activity of these animals and of spiders, G. Williams has formulated the following general rule: where a species occurs over a wide range of environments its nycthemeral periodicity is more sharply circumscribed in the more favourable habitats.[19]

Under natural conditions several external factors are probably active, at the same time modifying the expression of the circadian rhythm, generally with one in particular being the ruling factor of an animal's periodicity. Thus insect flight behaviour is influenced by a number of environmental stimuli. Light intensity is of course the most important of these, but the time of flight is also affected by season, day-length, wind and cloud-cover. Moonlight inhibits flight in Lepidoptera, but initiates activity in lower flies (Nematocera) above the minimum threshold temperature necessary for activity. High temperature may induce polyphasic flight in many nocturnal insects, while even day-flying species remain active till midnight during warm summer nights in the temperate regions of the world. High relative humidities (above ninety per cent) stimulate the flight of various Nematocera, Trichoptera and Miridae. Wet nights are most unfavourable to insect activity, although some Diptera are favoured by such conditions; heavy rainfall inhibits flight, as does high wind velocity. Larger insects tolerate higher wind speeds than do smaller species, but no flight occurs at speeds in excess of 14m. per sec.[20]

The nocturnal activity of flying insects, as shown by light-trap catches, is thus considerably modified by weather conditions. The minimum temperature, as well as the maximum of the previous day, are both higher on good nights when activity is maximal than on poor nights when it is slight. The wind is also calmer on good nights when the moon is closer to new than to full, and the barometer is high. Rain during the previous daytime is associated with a low catch, but rain during the night occurs as frequently with high catches as with low ones; relative humidity, likewise, does not appear to affect the catch.[21] A recent study of the lycaenid butterfly *Heodes vergaureae* has shown that activity in the field is largely dependent on air temperature, solar radiation and wind velocity. A low temperature can be compensated by high intensity of radiation and vice versa.[22]

A marked circadian periodicity of adult emergence is seen in *Trichopoda pennipes*, a species of tachinid fly that parasitizes the pentatomid bug *Nezara viridula*. This periodicity is greatly modified, however, by thermal changes, and adults can be induced to emerge at any time by small increases in temperature.[23] Most beetles of the family Carabidae are either nocturnal or day-active, but a few are plastic and their time of activity changes in response to variation in environmental conditions. This variation may be the difference between two adjacent habitats, between two parts of the geographical range of a species, or the difference in weather from day to day.[24]

Some of the concepts reached from the study of insect rhythms are also applicable to other taxa. Recent work on the plains garter snake (*Thamnophis radix*), for instance, suggests that in this species, and probably others, the crepuscular or nocturnal activity exhibited in hot weather may be caused, not by a direct response to high temperature or by temperature selection, but by the effect that the level of temperature has on the biological clock and on activity-controlling mechanisms.[25] Again, the threshold of panting in the Australian lizard (*Amphibolurus muricatus*) varies with the time of day, the midnight value being considerably lower than daytime values. The existence of circadian fluctuations in panting threshold is correlated with thermal changes in nature. During the hottest part of the day the panting threshold is high whereas at night, when excess heat is seldom a problem, it is at its lowest.[26] Similar cycles have been observed in the heat tolerances of frogs, while circadian differences in preferred temperature have been demonstrated in lizards and salamanders. They reflect a general shift in the entire syndrome of thermal responses in the animals concerned and are not merely responses to exogenous stimuli. In other words, the parameters of homeostasis vary in a circadian manner as we have already seen.

The biological clock enables many animals to emerge from their retreats accurately at the most propitious time. This has been demonstrated in a striking manner in the case of nocturnal mammals, such as flying foxes and bats, which are woken through the action of their endogenous rhythms and emerge promptly at sunset.[27] Investigation of the environmental synchronizers of the circadian rhythm of squirrel monkeys (*Saimiri sciureus*) has shown that even in the absence of other time cues, only light–dark cycles and cycles of food availability are entraining agents.[28] Weather is therefore unlikely to affect the circadian rhythms of these animals directly, although it may be significant that thermoregulation is impaired in an environment without circadian time cues. Exposure to cold produces a significant decrease in core body temperature when the circadian rhythms of the animal are free-running in the absence of environmental synchronizers.[29]

It is a point of considerable interest that the circadian nature of the biological clock is, itself, functionally significant. It is not a tolerated error, but an integral part of the entrainment mechanism whereby the system can assume the frequency of the environmental cycle, and do so with a determinate phase relation to it that is adaptively appropriate.[30] The deviation of the free-running period from that of the solar day is astonishingly constant. It is more than enough to account for the precision with which animals are able to compensate for the movement of the sun, as we shall see in chapter 6. Furthermore, it seems probable that the extent of the deviation may be related to the day-length at the latitude of an animal's natural habitat. Thus the mechanism of biological time measurement, like that of man-made clocks and watches, contains built-in controls that can be used for adjustment and synchronization.

Chapter 5

Rhythms of Feeding, Predation and Dispersal

IN NATURE, as in everyday human life, there are both good times and bad times for carrying out normal activities. Pliny wrote that to digest food while sleeping is more conducive to corpulence than to strength.[1] Modern research on circadian rhythms indicates that in general digestion is not much affected by sleep; with respect to particular secretory and motor phenomena, however, especially of gastric function, there is no universal agreement among investigators.[2] On the other hand, there is no doubt that the times at which animals normally feed are closely bound up with the availability of their food. For instance, carnivorous animals with a limited variety of prey must show a periodism that is closely correlated with that of the prey. Larger carnivores, with a wider choice, are less specialized with regard to their hours of feeding than are smaller forms.

The times at which animals are most active depend upon a number of environmental factors, both physical and biotic. The latter include both predator pressure and the availability of prey, while quality of vision is often adapted to rhythm of activity, as we saw in the previous chapter.

Feeding rhythms of invertebrates

The feeding rhythms of smaller animals, like their rhythms of locomotion, are often controlled by the physical conditions of the environment. The activity of earthworms is restricted to the hours of darkness when the temperature falls and the relative humidity of the atmosphere increases. At the same time, activity and feeding are often interrelated. Investigations of a natural infestation of cockroaches and bed-bugs have indicated that cockroaches are active just after sunset, while bed-bugs show peaks of activity just before dawn, every three to four days. These natural cycles continue for several days in total darkness, but only the bed-bugs come out in constant light. The endogenous clocks in insects give a stimulus which may or may not be acted upon, depending on other stimuli. The feeding rhythm of bed-bugs is controlled by temperature: a bug at 30°C feeds every three days, at 20°C every six days, whereas the clocks which control the actual time on the third or sixth day at which they feed are independent of temperature.[3, 4]

More investigations have been carried out on the periodicity of feeding, among blood-sucking flies, than on any other aspect of the circadian rhythms of insects. In particular, the important research of A. J. Haddow, J. D. Gillett, P. S. Corbet and their colleagues in Uganda has shown that the nycthemeral biting-cycles of mosquitoes and other flies are related to microclimatic conditions, especially of saturation deficiency, and that the rhythmic vertical movements of the insects are controlled by them.[5] From a microclimatic point of view, the canopy layer of branches and leaves during the day is similar to the forest margin near the ground. At night, however, there is no significant difference between the various levels and no physical barrier to inhibit the diffusion of mosquitoes from the forest into the atmosphere above and around it. The biting-cycle results from a combination of a true rhythm of biting with one of flight activity, which leads to a concentration of mosquitoes in the neighbourhood of man and other host animals.[5] This is followed by a period of quiescence, before a further approach is made to the host.[6]

The circadian rhythm is not the same in all species and the time of activity may be influenced by the environment. Exogenous factors may act directly on insects, triggering a change in activity as, for example, when an insect resting or engaged in non-directed flight is induced to follow some other special activity, such as seeking a blood meal or swarming. Light intensity may operate either as a direct stimulus, or as an inhibiting factor whose removal is the stimulus-inducing activity.[7] The behaviour of mosquitoes in nature results from the complex interplay of exogenous and endogenous influences. Light, temperature, humidity and wind all have their effect. The circadian rhythm not only controls the time at which a mosquito takes flight, but determines the duration of the flight period.[8] Thus the feeding-cycles of mosquitoes, as of other insects, are mainly an adaptation to physical and meteorological conditions of the environment.

In the case of bees, as of sphingid and noctuid moths, the times of opening of the flowers which provide their sources of food are obviously an important factor; however, other factors may in the past have determined the time of activity of the moths, to which the flowers themselves have become adapted. The evolutionary priority of cause and effect cannot therefore be determined.[9]

The circadian rhythm of an African grasshopper (*Poecilocerus hieroglyphicus*) is synchronized primarily by alternating light and darkness, while the daily patterns of feeding, mating and oviposition are correlated with locomotory activity. The number of insects feeding reaches a peak after sunrise and falls to a minimum during the night.[10] A possible ecological explanation may lie in the fact that the grasshoppers are brightly coloured and poisonous to predators. It is important, therefore, that their activities should take place at a time when they are easily seen and avoided.[11]

In the last chapter, we saw that the carabid beetle (*F. madida*) is nocturnal in woodland where there is a rich fauna at night but day-active in open country. According to H. Remmert, synchrony of diurnal activity patterns among animals seems to have evolved entirely between groups of species, and no well-established case of synchrony is known that involves only two species.[12] The interdependence of activity patterns, based on circadian rhythms, is a phenomenon well known in ecology. Examples include the interactions between flowers and the insects that pollinate them, parasites and their hosts, predators and their prey. At different times of day completely different food chains can be distinguished in the same biotope. The few quantitative investigations which have been carried out reveal that strong selective pressure can limit the diurnal activity of a species, and that the productivity of a biotope may reach a maximum when the daily feeding time of its predators is restricted, as in the case of marine plankton.

Vertebrate feeding rhythms

The daily activities of animals tend to be extremely regular. This is not really surprising, however, since most innate behaviour is incapable of modification. Some animals are active during a certain period of the day or night, but others exhibit varying kinds of activity at different times.

The locomotor activity of West Indian fossorial amphisbaenians and worm snakes, for instance, shows evidence of strong endogenous circadian rhythms in continuous light. In addition to avoidance of high daytime temperatures, it is believed that nocturnalism may be associated with the time of activity of the prey, such as burrowing invertebrates and army ants, which are also nocturnal.[13]

Many normally day-active birds migrate at night, while aquatic insects which spend the daytime swimming about in ponds and streams usually fly from one locality to another during the hours of darkness. An interesting example of the periodic movement of animals is recorded from a small island off the coast of Sri Lanka. This is inhabited nightly by crows which roost in the trees and fly to the mainland in the morning to feed. During the day their place is taken by fruit-bats or flying foxes which fly to the island at dawn.

Under normal conditions most birds have fixed drinking patterns, but they may be temporarily affected by disturbances such as those caused by other animals coming to drink, the presence of birds of prey, and so on. Birds that normally drink only once daily, such as sand grouse, doves, and other desert species, usually drink during the early hours of the morning. Apparently they need water after the long night's rest.[14]

At first sight it might appear that feeding rhythms ought to be more marked among carnivores than in herbivorous animals whose food is always present. The activities of herbivores, however, may be related to escape from enemies, as we shall see, and therefore rigidly selected on a temporal basis. Moreover, there may be distinct nutritional advantages in feeding at night. The moisture content of tree leaves, upon which many desert browsers depend for water, is much higher at night than during the day. For this reason, gazelles, oryx, eland and other antelope benefit considerably from their nocturnal feeding patterns and are far more efficient as producers of protein in arid regions than are domesticated cattle which normally feed during the hours of daylight.

Mammalian carnivores are difficult animals to study because they are highly adaptable and secretive and, in captivity, often need to be fed daily. In some cases the taking of a large meal influences subsequent activity markedly. As the result of a study based on locomotory activity, measured by means of running-wheels, it would appear that such animals can be divided into a number of groups. Ringtails (*Bassaricus astutus*), kinkajous (*Potos flavus*) and genets (*Genetta genetta*) are stictly nocturnal. The behaviour of kitfoxes (*Vulpes macrotis*), long-tailed weasels (*Mustela frenata*), and red foxes (*Vulpes vulpes*) suggests visual systems best adapted for dim light, but also suitable for daylight, while the arctic fox (*Alopex lagopus*), grey fox (*Urocyon cinereoargenteus*) and striped skunk (*Mephitis mephitis*) also appear to be adapted for both day and night vision. Finally, animals adapted best for daylight vision include mongooses (*Herpestes auropunctatus*), coatis (*Nasua nasua*), tayras (*Eira barbara*) and the grison (*Galactis vittatus*), but the last three also appear to be well-suited for vision in dim light. The coyote (*Canis latraea*), grey fox, red fox, long-tailed weasel, mongoose and red wolf (*Canis niger*) are markedly inhibited by darkness, but the coati, genet, kinkajou, kitfox, ringtail and skunk are uninhibited, or only partly inhibited.[15, 16] These conclusions are supported by published observations on animals in nature.

In a study of the activity patterns of eight species of grassland rodents belonging to the family Muridae, all were found to be either nocturnal or crepuscular in habit. This information, combined with their food and habitat preferences, provides an insight into the ways in which these animals exploit their habitats. It was suggested that humidity may be a significant factor in determining activity, particularly in an environment where water conservation is important. Thus, of the two most common species, the punctuated grass mouse (*Lemnoniscus striatus*) is crepuscular whilst the multimammate rat (*Praomys natalensis*) is nocturnal. The latter feeds on insects, especially termites, throughout the year whereas in the crepuscular species, termites form an important part of the diet only during wetter months.[17]

The tendency for many mammals to be active at dawn and dusk probably hinges on these periods' being the optimal times for hunting and food gathering. Twilight is often a time of moderate temperature and the only period in which the activities of many species overlap. It is therefore generally the best time for small carnivores to hunt other small nocturnal mammals, not only because the latter are usually intensely active at these times but also because visibility is better at night, and the relatively dim light provides partial protective cover for the small carnivores from the larger predators such as hawks and eagles. The fact that many insects are crepuscular and therefore most active at twilight is an influential factor in timing the activity of rodents and small carnivores.[18]

No doubt many animals are crepuscular because it is advantageous to utilize all the available time that is suitable. This would be particularly true of nocturnal animals, for atmospheric refraction shortens the night, and the shortest nights occur during the season in which food is most abundant and the hoarding of food by rodents at a peak. Even nocturnal animals sometimes omit dawn activity because hunting and food gathering at dusk and during the night frequently provide them with adequate supplies.

Twilight is thus generally the best period for small and medium-sized carnivores to hunt small nocturnal prey. Not only are the latter intensively active at this time, but visibility is better than at night, while the dim light provides protection from larger carnivores. In general, therefore, small animals do not usually emerge until lighting conditions tip survival factors in their favour, and they probably retire before the light is sufficiently strong to assist their predators unduly. The interaction in twilight between predators and prey may well account for the remarkable correlations between light intensity and the speed of running that have been established experimentally. As light intensity increases so does the velocity of locomotion, for natural selection would obviously tend to favour animals that adjust their speed of movement to allow for the increasing brightness of the light.

Escape from enemies

During their periods of inactivity, many animals shelter in holes, burrows and retreats where they can avoid unwelcome predators as well as unfavourable climatic conditions. These burrows may be closed by a plug of silk, earth or fallen leaves. In the case of terrestrial snails, the opening of the shell is occluded by an epiphragm of dried mucus. Phragmosis, in which the burrow is closed by a part of its occupant's body, is an extreme example. It is exhibited most notably by certain ants and termites, where the head of a soldier ant is adapted to fit the openings in woody plants that these insects inhabit. The name 'phragmosis' was coined by the late W. M. Wheeler from the Greek. It means a fence or barricade. A truncated head for closing the entrance to the nest has been found in various ants and termites. The same strategy is employed by the larvae of certain tiger-beetles (Cicindelidae). An unusual example of phragmosis is found in a spider (*Chorizops loricatus*) of tropical America, the end of whose abdomen fits, like a cork, into the burrow entrance whenever the spider is in residence.

The harvester termite (*Hodotermes mossambicus*) escapes predation by matching its spatial and temporal activity to be the inverse of that of its enemies. In one region of South America, where birds are the major predators and are mainly active in the morning and late

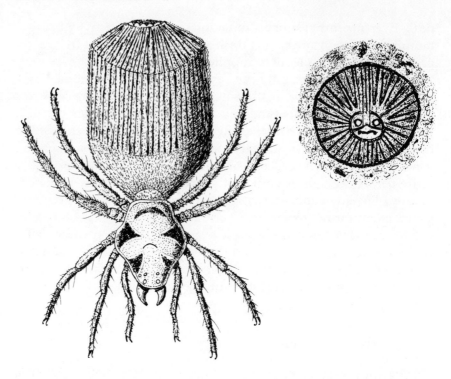

16. *Chorizops loricatus*, a spider of tropical America, the end of whose abdomen fits like a cork into the entrance of its burrow, thus protecting the spider from enemies and maintaining a favourable microclimate within. Left: seen from above showing truncated end of abdomen. Right: viewed from the rear. (After W. M. Wheeler.)

afternoons, termites show peaks of activity in the middle of the day and during the night. In another area, where enemies include both birds and predatory arthropods – mainly spiders – the activity rhythms of the termites show a single large peak in the hottest period of the day. It seems that their spider predators are compelled to be active at this unfavourable time also, since there is no other suitable prey available for them. When predation occurs news of the event is communicated by means of chemical alarm substances, or pheromones, to all members of the termite colony, and result in a well-organized evacuation of the area in the vicinity of the predator.[19]

Another way in which the effects of predation can be reduced is by synchronization of emergence times so that the appetites of the predators are rapidly satiated and they are unable to consume more than a small fraction of the total food available. The adaptive significance of bats emerging from their roosts each evening at the same time in relation to sunset probably lies in the avoidance of predators such as kites and the bat-hawk (*Machaërhamphus anderssoni*). This can catch its prey only at dusk and dawn when the bats are active and there is yet sufficient light for them to be seen. No other food is taken. In Khartoum, there is often a brief overlap between the roosting of the kites (*Milvus migrans*) and the emergence of the bats. During this period the earliest bats are sometimes caught by the last kites retiring for the night.[11]

Depending on the environment, synchronization of emergence may increase or reduce predation on insects. Thus the mass emergence of dragonflies along the river bank may attract large numbers of territorial birds to feed on them; but if a pair of birds successfully defends a territory which includes an emergence site, mass emergence reduces mortality, because more individuals emerge than a single pair of birds can eat.[20] Synchronized emergence of aquatic insects, as an adaptation to predation, occurs more frequently in consequence of circalunadian than of circadian rhythms. Synchronized avian breeding cycles perform a similar function.

A census of game animals carried out at water-holes in Rhodesia has provided data on the daily drinking habits of the more common herbivores and carnivores. Populations of herbivores fall into two groups: buffalo, zebra, elephant and giraffe which drink in the evening and at night; and wildebeest, roan, sable and kudu which drink during the night and in the morning. It may be surprising that the commonest, largest and most aggressive species should all drink together in the evening. Individual animals, however, with the exception of wart-hogs which drink exclusively during the day, have been recorded drinking at most hours of the day and night. Whereas lions predominate at drinking-pans during the day, leopard are usually recorded from just before dusk to just after dawn, while hyaenas are more exclusively nocturnal.[21] Drinking is a dangerous occupation at any time, but the risk of predation is probably reduced when a number of animals are present so that they can give each other warning of the approach of an enemy.

Avoidance of competition

Competition is one of the most intense selective factors in ecology. It operates between sympatric species – species which inhabit similar environments – as well as between members of the same species (intra-specific competition). Many animals escape competition for food, or gain protection from enemies, by assuming a nocturnal activity period.[22] Not only do more primitive, and therefore presumably less efficient, orders contain a higher proportion of nocturnal forms than do more advanced orders of insects, but the more primitive species of higher orders also tend to be active at night.[23] Whether this generalization can be applied to higher animals, such as mammals, is debatable since even among invertebrates a number of highly specialized forms are nocturnal. When these specializations are considered they are seen to consist chiefly of refinements in the senses and are of degree rather than of kind. Indeed, luminescence and some adaptation of vision are the only examples of modifications not equally useful by day.[24]

Although competition may be one of the most important biotic factors influencing the timing of circadian rhythms, it tends to be a variable one. Indeed, studies of the physiology of vertebrate eyes which, as we saw in the last chapter, may become adapted to daytime or night vision, suggest that diurnality and nocturnality come and go as mutation and ecological expedience direct.

The nocturnal habit confers several advantages. Enemies are more easily avoided and food more easily obtained than during the day. Losses of moisture are reduced, and olfactory communication between members of the same species is greatly facilitated at night because odours are more readily conveyed when there is no solar radiation to create currents and eddies in the atmosphere. Perhaps primitive and less efficient species become nocturnal in habit as a result of competition with more advanced and efficient types. In view of the difficulty of defining what is really meant by the word primitive, however, and in the absence of any quantitative data, it will probably be wise at present to maintain an open mind on the question.

Strictly nocturnal birds and mammals tend to be larger than day-active species, and have zoogeographical affinities quite different from those of diurnal forms occupying the same areas. Predators may become secondarily nocturnal and capture their prey at a time when it is least able to defend itself, while many herbivores benefit from nocturnal habits, thereby avoiding water loss, overheating and the attentions of blood-sucking insects.[25] Not only are many factors involved, but they often influence one another. Some nocturnal animals are never active in the daytime while others can alter their habits in different circumstances. Thus the African buffalo was very abundant until the terrible rinderpest epidemic of the 1890s exterminated the animals in many places. Whereas buffalo used previously to feed in herds in the open by day, the survivors retired to forests and dense swamps, feeding only at night. When their numbers recovered buffalo returned to their former diurnal behaviour. Game animals, likewise, tend to nocturnal habits when much hunted, and rabbits do not come out until nightfall in areas that are often shot over. Another example is provided by the behaviour of the Nile crocodile in Lake Victoria. Where they have been intensively hunted, these fine reptiles have become increasingly nocturnal and are quite commonly encountered by night in places where none can be seen during the day. In nearby game parks, however, where crocodiles are not molested, they can regularly be seen by day as well as at night.[9]

It has recently been found that, contrary to most published accounts, bushbuck (*Tragelaphus scriptus*) are no more active at night than during the day but exhibit cycles of activity with a periodicity of two to five hours. A crepuscular peak of movement is, however, apparent. Bushbuck move out of thickets into grassland at dusk, graze mainly at night and return to the thickets at dawn. It can be argued that while association with thickets may be of advantage in avoiding predators that rely on vision during the day – chiefly lions and men – such an advantage is less important at night. Association with thickets may even be disadvantageous then, since it would greatly limit the area in which a potential predator – a lion, hyaena or leopard – would need to hunt. Thus while the distribution of bushes sharply limits the distribution of bushbuck during the day, this limitation becomes less significant at night.[26]

In equable climates, in which the environment is most predictable and suitable for activity throughout the twenty-four hours, there is a particularly high adaptive value in possessing a rhythm which can reduce inter-specific competition by temporal diversification. Insects, in particular, tend to be opportunist in such matters. For instance, light-trap catches of male horse-flies (*Chrysops centuriones* and *Tabanus thoracinus*) in Uganda have revealed a biphasic activity rhythm in open woodland with peaks at dusk and dawn, whereas in forests activity is restricted to a single hour before sunrise.[27]

Closely related species living in the same habitat frequently show differences in their time of activity. A good example is afforded by army ants of the genus *Dorylus*. Although males fly throughout the year in Uganda, analysis of light-trap catches shows that each species enters the trap at a particular time of night. The peaks of entry form a remarkably regular series from sunset until just before dawn. It is not easy to understand what the significance of this succession can be, unless it is that it results in the avoidance of competition between congeneric species; it is difficult to imagine that so marked a phenomenon can be without adaptive value.[28]

Rhythmic colour-change

In his *Historia Animalium*, Aristotle remarked that when the octopus comes into the neighbourhood of small fishes it may change its colour so as to resemble the surrounding stones, and that it reacts in a similar way when it is alarmed. These comments contain the essential elements of the modern theory of crypsis, or concealment, for offensive and defensive purposes. Similar observations have subsequently been repeated from time to time by other naturalists. Although many anti-predator devices cannot be adapted to suit particular circumstances, cryptic coloration, when achieved by means of chromatophores or pigment cells, enables animals to change their hues to match the background when they move from one environment to another. Examples are found among numerous crustaceans, cephalopod molluscs, fishes, amphibians and reptiles, which may show concealment in defence and offence or adopt conspicuous coloration for purposes of display during courtship and so on. Colour-changes, which result in crypsis and must therefore be mediated through the eyes, naturally occur at dusk and dawn as a direct response to changes in illumination. At the same time many cases of colour-change are under control of the biological clock.

In addition to their well-known tidal rhythms, fiddler crabs (*Uca* spp.) exhibit a number of adaptations to the twenty-four-hour day–night cycle. The elaborate displays of the males, and their accompanying pigmentary changes, are of functional significance only in daylight when they can be perceived by other crabs. Jocelyn Crane has observed that tropical fiddler crabs are largely day-active, and include among their members those forms in which display and colour-change have reached their highest development. Furthermore, she has described an apparent relationship between the phylogenetic position of a species and the time of day at which it reaches the maximum intensity of display. The period of display of primitive species is often restricted to the hours immediately after sunrise, while that of more advanced forms is displaced to a later time of the morning.[29]

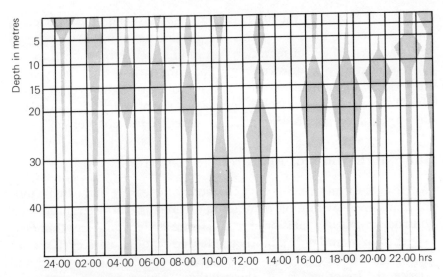

17. Daily migration of a small crustacean (*Cyclops strenuus*) in Lake Windermere. In the construction of the figure, the catch at each depth has been converted to a percentage of the total catch. (After P. Ullyott.)

In contrast to tropical fiddler crabs temperate species are active at low tides whether they occur during the day or at night. The experimental analysis of their timing systems has been carried out by F. A. Brown, Milton Fingerman, F. H. Barnwell and their colleagues.[30, 31] Endogenous rhythms of locomotory activity, oxygen consumption and colour-change have both tidal and circadian frequency. Because the chromatophores remain contracted during the night, the tidal rhythm can be clearly expressed only in daytime when the pigment is dispersed. Tide-related variations in the daily pattern are reproduced at fifteen-day intervals in accordance with the semi-monthly repetition of the tidal cycle, and this is timed by the interaction of the daily and semi-diurnal tidal rhythms.

Adaptive colour-changes such as this are found in many shore-dwelling animals and can have several functions, although crypsis is probably the most usual. Similar responses are known in terrestrial forms, especially reptiles, of which the chameleon is a well-known example. They may be under exogenous or endogenous control.

Other physiological responses to the light and dark cycle which may have an endogenous basis include the nycthemeral migration of the eye pigments in many crustaceans and insects. By this means the eyes become automatically adjusted to light, even when the animals are maintained in constant darkness. An endogenous rhythm of flashing is shown by the firefly (*Photinus spiralis*) and other species, as well as of light production in luminescent marine copepod crustaceans. Many observations on luminescence have been made. Although it may recur regularly in constant darkness, in every case it is noticeable that light production is inhibited by light, even at night. In all these examples it would appear that the primary function of the biological clock is preadaptation, as discussed in chapter 4.

Diurnal rhythms of migration in plankton

Vertical daily movements are a characteristic feature of the behaviour of many free-swimming planktonic organisms. In different species diurnal migration may be due to strong endogenous rhythmicity, synchronized and only slightly reinforced by light–dark cycles; or to weak rhythmicity, strongly reinforced by the cycle of daylight and darkness. In other cases, the rhythm appears to be exogenous, depending upon direct reactions to light – simple phototaxis and kinesis.[32] The phenomenon is well-documented, and has been reviewed on many occasions, but its adaptive significance is not well understood.[33]

In general, most planktonic species avoid strong light, each showing a preference for a certain intensity. For this reason few organisms are to be found in the surface layers of the sea during the hours of daylight; they are distributed at various depths according to their specific light responses. At dusk they tend to swim upwards, but when all is dark and there is no light stimulus, they tend to scatter. They migrate to the surface again at daybreak and later move downwards as the light strengthens. Doubtless the plant or phytoplankton tends to aggregate in the light intensity optimum for photosynthesis, while the reactions of the animals tend to maintain them in regions where they find a rich supply of food. But this is not the whole explanation. There is often an inverse relationship between the distribution of the plant and animal plankton.

Diurnal migrations occur even among deep-water animals, and the movements of the species most abundant in the Sargasso Sea, for example, are obviously related to day and night. Moreover, the lower limit of the penetration of daylight of an intensity significant in determining the vertical distribution of animals (about a thousand metres in clear water) is correlated also with marked changes in temperature, salinity, oxygen and phosphate content. It has therefore been suggested that a combination of factors may be responsible for keeping the majority of the macro-plankton above a thousand metres in these waters.

The various factors of the environment controlling diurnal migration of planktonic animals have been listed in the following order: (a) light, which clearly dominates under average conditions; (b) temperature, which becomes very important and can even overwhelm the effect of light when it exceeds 20°C; (c) other factors such as salinity, aeration and so on. The actual mechanism by which organisms keep themselves at the optimum level may be responses to light, gravity and the acceleration or inhibition of movement. It may be a combination of them all and, from the evidence of laboratory experiments, would seem to vary in different species.

The diurnal cycle of vertical migration in planktonic Crustacea has been shown to consist of four parts: ascent from the daytime depth, midnight sinking, dawn rise, and descent to the daytime depth. Ascent in the evening and descent in the morning are related to decreasing and increasing penetration of daylight. The midnight sinking is probably the result of the passive state of the organisms in full darkness while the dawn rise represents a return by the animals to optimum light intensity. This is supported by the fact that the order in time of arrival at the surface of some freshwater species is the same as the order in depth at which they are to be found in full daylight. Migration between periods of complete darkness may be occasioned by directed (phototaxis) or undirected (photokinesis) response to light, or by a combination of these mechanisms.

It has been shown that a population of water-fleas (*Daphnia magna*) in a tank filled with a suspension of indian ink in tap water, will undergo a complete cycle of vertical migration. A 'dawn rise' to the surface at low light intensity is followed by a descent of the animals to a characteristic maximum depth. The animals rise to the surface again as the light decreases and finally show a typical midnight sinking. The light intensities at the level of the animals are of the same order as those recorded in field observations.[34]

Optimum light intensity may be an important factor in influencing the vertical migration of planktonic organisms, but a number of other factors are also involved. Some animals which usually stay down in the daytime may occasionally be seen at the surface in bright sunshine. It is also clear from the result of tow-netting at different levels that all individuals in a population do not react in the same way to one particular set of conditions. Even though the majority may migrate upwards, a proportion usually remains below.

A. C. Hardy discussed the matter in some detail, and suggested that vertical migration might have been evolved because it gives the animal concerned a continual change of environment which would otherwise be unattainable for a passively drifting creature. Water masses hardly ever move at the same speed at different depths, for the surface areas are nearly always travelling faster than the lower layers. So, although by swimming in a horizontal direction an animal will not get much change of environment in the sea, by moving upwards and downwards it may achieve an extensive degree of change. The same author also proposes a hypothesis of animal exlusion to account for some of the observed phenomena of plankton migration. According to this, the distributional relationship between animals and plants may be due to a modification in the vertical migration of the former in relation to dense concentrations of the latter. Most animals, it is supposed, would come up to feed in the dense phytoplankton zones for a short time only, possibly because of some antibiotic effect, and would therefore tend to become distributed in larger numbers in other areas. On the other hand, the animals which stay up longest in the phytoplankton would naturally remain more concentrated in these regions. This explanation is largely hypothetical, although certain flagellates are known to have a poisonous effect.[35]

Alternatively, animals may become concentrated by modified vertical migration in regions rich in phytoplankton, graze there for a while, and then move to other areas by vertical migration. That the vertical diurnal migration of plankton has important adaptive functions is clear from its ubiquity, but the exact nature and relative importance of these have yet to be assessed.

Hardy's idea was subsequently extended by the suggestion that marine planktonic species would tend to become divided into small, separate populations if they were to drift continuously in one stratum, since they would not normally encounter any directional stimuli to horizontal migration. Gene flow is promoted throughout the population, however, by species migrating at different times to different depths where the water layers show varying speeds of current. Although no conclusive answer can be given to the question as to the adaptive function of vertical migrations, the extensive occurrence of the phenomenon clearly suggests that it is of great significance to planktonic organisms. It is not improbable that the most important benefits derived by one species or group of organisms will differ from those derived by other species or groups.[36]

Probably there is some truth in each of the hypotheses that have been postulated to explain the significance of vertical migration. No doubt this type of behaviour in a three-dimensional environment provides the best chances of utilizing to the full the resources of that environment. Different species are separated both in time and, frequently, in space, because the movements of each species are related to those of different water masses.[37]

Nycthemeral rhythms of locomotory activity on land have been considered in relation to climatic conditions and to feeding. Although there is no parallel, in an essentially two-dimensional terrestrial surface, to the dispersal of planktonic forms resulting from diurnal vertical migrations, there is quite a close analogy between conditions in the sea and in the air. It is by no means impossible that nycthemeral rhythms of flight activity in insects may serve to enhance the dispersive effects of locomotory movements, but this hypothesis has never been tested quantitatively. Nevertheless the uses, both direct and indirect, to which animals and plants put their biological clocks are so diverse that the biologist is never surprised when he comes upon unexpected or bizarre examples. The measurement of time is not only important in regulating feeding, predation and dispersal by animals but also in determining the time and direction in which they migrate.

Chapter 6

Animal Navigation

THE TWO main parameters of animal migration are both determined by biological clocks. The first of them, celestial navigation, often depends upon the accurate measurement of time. It is discussed in the present chapter. The second, seasonal timing of migration, is usually achieved through the measurement of photoperiod by means of the circadian clock, and will be discussed in chapter 8.

An ability to steer by the position of the sun and to analyse the pattern of polarized light in the sky enables many arthropods to direct their movements. Similarly birds, fishes, and other vertebrates are known to orientate their movements by observation of the sun's position. Since the direction taken remains unchanged throughout the day, it follows that allowance must be made for the apparent movement of the sun across the sky between dawn and dusk. In other words, solar or sun-compass navigation is time-compensated.

Time-sense of honey-bees

The ability of honey-bees (*Apis mellifera*) to return to a source of food at the same time each day has been known since the beginning of the present century, when it was observed by H. von Buttel-Reepen that bees only visited a field of buckwheat in the morning when the blossoms were secreting nectar. A few years later, the Swiss naturalist August Forel who, with his family, often had meals on a terrace in their garden, noticed that honey-bees would appear at the table at meal-times, whether or not food was actually present. Forel therefore suggested that bees might be able to estimate the time by observation of the sun, but he did not carry out any experiments to determine whether or not this was the case. Indeed, no further observations of any kind were made for over twenty years but, more recently, a considerable amount of research has been carried out on the time-sense of honey-bees.[1]

Bees have been trained to visit the same feeding place not merely once every day, but even up to nine different times, and to distinguish between two points in time that were not more than twenty minutes apart.[2] The timing of such visits is not affected by temperature or weather, and training is not disturbed when the experiments are conducted under constant conditions in the laboratory. Indeed time perception appears to be innate, since bees have been trained which have hatched in a dark chamber and have never experienced the alternation of day and night.[3] The time-sense is not based on a learning of intervals because many honey-bees cannot be trained to intervals deviating from a twenty-four-hour periodicity, such as an interval of nineteen, twenty-seven or forty-eight hours. It is noteworthy, too, that they always visit the right place at a given training time.

Despite conflicting opinions, it now seems generally agreed that increasing the rate of metabolism with thyroxine, or decreasing it with quinine or other drugs, has no effect upon the time-sense of the honey-bee, but cooling to 5°C for about five hours causes a three- to five-hour delay in the time of their arrival at test feeding dishes. These and other experiments indicate that the time memory of bees depends upon an endogenous clock which possesses properties common to all circadian systems. It is a continuously consulted clock, with clear affinities to an activity rhythm, and its adaptive significance is obvious. It enables bees to return to a known food source when nectar and pollen are most readily available, and therefore to achieve maximum productivity. Moreover, if the bees are compelled to remain in the hive for a couple of days because of bad weather, they can still remember when a particular plant secretes its nectar. At the same time, the rhythm is fairly easily extinguished without positive reinforcement so that plants are not visited after they have ceased flowering, and other sources of food are remembered in their place.[4]

Bees are convenient experimental animals and their biology has been investigated in considerable detail, but they are by no means unique in possessing a sense of time. Many unrelated insects are able to learn the times at which different kinds of flowers offer their nectar and pollen, while they can also use their circadian clocks, as do other species, for time-compensated celestial navigation.

Orientation to the plane of polarized light

In 1788, M. J. E. Spitzner[5] wrote 'When a bee finds a good source of honey somewhere, after her return home she makes this known to the others in a remarkable way. Full of joy, she waltzes around them in circles, without doubt in order that they shall notice the smell of honey which has attached itself to her; then when she goes out again they soon follow her in crowds.' This observation was forgotten until 1920, when K. von Frisch published his first paper on communication between bees by means of dances.[6] The story of the bees' dance is now well known, even if its significance still remains somewhat controversial.[7] It lies beyond the scope of the present volume, and I have only mentioned it because von Frisch's work on the communication by bees of the direction of food crops also revealed that honey-bees can appreciate the plane of polarization of light. Light travels in waves which vibrate transversely, that is, at right angles to the direction of travel. In ordinary light these vibrations lie in an infinite number of transverse planes but, in polarized light, they are in only one transverse plane. The light reflected from any part of the blue sky is partly polarized, the plane and the degree of polarization depending on its direction in relation to that of the sun.

The human eye cannot distinguish between ordinary and polarized light, but arthropods can even distinguish the direction of the vibrations, a facility they make use of in their orientation. For this reason honey-bees are in no way disorientated when the sun is obscured by cloud. So long as a part of the sky remains visible they are able to maintain their sense of direction.

Time-compensated solar navigation in arthropods

An ability to analyse the pattern of polarized light in the sky enables honey-bees to find their way to and from their sources of food. Similar principles govern the direction of movement of many other kinds of arthropods. It has been known since 1911 that black lawn ants (*Lasius niger*) use the position of the sun, the light compass reaction, to maintain a straight course in territory poor in landmarks. If shielded from the direct rays of the sun but exposed to its reflection in a mirror, they alter their course as though steering by the reflection of the sun.[8] Earlier work had suggested that if ants were detained for a few hours in a darkened box and then released, they would continue at the same angle to the sun and, therefore, in a different direction. More recently, however, compensation for the changing azimuth (the horizontal angle) of the sun has been described in black lawn ants and, also, in wood ants (*Formica rufa*) but only during the summer. In March and April, the latter species is apparently unable to compensate for solar movements and shows an incorrect compass direction after being imprisoned in the dark. Apparently compensation for the changing azimuth has to be learned anew after the winter.[4]

Time-compensated sun compass orientation has also been described in locusts, beetles (*Geotrupes sylvaticus*), pond skaters (*Velia currens*), sand-hoppers (*Talitrus saltator*, *Talorchestia deshayesei* and *Orchestia mediterranea*), wolf-spiders (*Arctosa perita*, *A. cinerea* and *Lycosa fluviatilis*) and other arthropods. It is known, for instance, that various littoral sand-hoppers are able to return directly towards the sea if moved inland and placed on dry sand. In some of the original experiments, sand-hoppers were transported from the west coast of Italy to the Adriatic shore. On release, they continued to move westwards, despite the fact that the nearest sea now lay to the east.[9]

The clock of the sand-hopper, like that of honey-bees and other insects, is very resistant to other environmental changes. When sand-

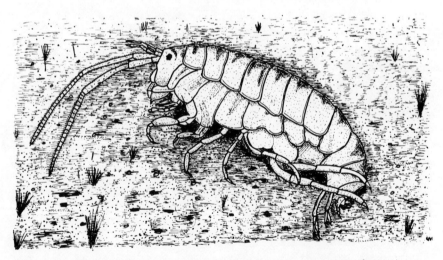

18. A sand-hopper. When moved inland these littoral crustaceans always move in the direction of the sea. Their sense of direction is based on observation of the sun or moon and their sense of time enables them to allow for the movement of either across the sky.

hoppers were taken by plane from Italy to South America, they orientated at an angle to the sun that would have been correct had they remained in Europe. The importance to a small crustacean, that would soon become desiccated in dry surroundings, of finding its way directly back to the damp sand of the seashore when accidentally transported inland or blown there by the wind, is quite obvious. Animals of each population must, however, learn a different direction according to the position of the sea in relation to the beach on which they live.

A similar type of orientation has been found in the wolf-spider *A. perita*. This animal often lives on the shores of lakes and rivers. If placed on the water, it hurries back to the bank in a direction perpendicular to the shoreline. But if it is taken to the opposite bank and placed on the water there, then it tries to cross to the shore on which it normally lives. It runs in the direction to which it is adjusted by the position of the sun. That the angle toward the sun must play a decisive part has been demonstrated by experiments in which animals were, like the ants cited above, misled about the position of the sun by means of a mirror. The participation of the time-sense has been demonstrated by experiments in which wolf-spiders were temporarily kept in darkness. The readjustment of these animals when shifted from one lighting regime to another takes about as long

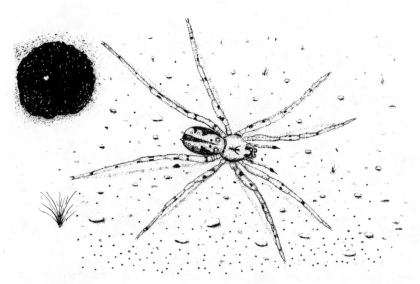

19. A wolf-spider at the entrance to its burrow. Wolf-spiders, like sand-hoppers, are able to maintain their direction by time-compensated solar navigation.

as re-synchronization of the circadian rhythms of locomotory activity would take. Phase shifting can be achieved by exposing the spiders to temperatures between 2°C and 5°C.[10]

The ryhthm of solar orientation has been shown experimentally to operate even during the night. The animals under investigation take up angles of orientation which alter in a regular manner between the angle appropriate to dusk and that appropriate to dawn. This nocturnal variation can, however, appear in two different patterns. In the first, which is found in sand-hoppers, beetles (*Phaleria provincialis*) and wolf-spiders (*A. variana*), the angle of orientation alters during the night by passing in reverse order through the values shown during the day, in such a way that the animals would be correctly orientated during the night if exposed to an hypothetical sun following the opposite course to that followed by the real sun during the day. In the second pattern, found in honey-bees, the angle of orientation alters during the night through new values which are not taken up during the daytime. The animals seem to calculate a complete revolution of the sun along the horizon and thus compensate for a solar movement through 360° in twenty-four hours.[11]

Italian populations of wolf-spiders (*A. cinerea*) are incapable of definite orientation during the night but Finnish populations of the same species, living within the Arctic Circle, orientate themselves correctly throughout the twenty-four hours, on the days around the summer solstice. Between sunset and dawn, the spiders orientate themselves by the polarized light of the sky.[11] No doubt the circadian rhythm is synchronized by daily fluctuations in light intensity. During the arctic summer, when there is continual daylight, it would be important for shore-dwelling spiders to orientate themselves throughout the time that they are active and during which they run the risk of being blown into the water.

Like insects, fishes are able to use the sun's position as a directional reference, and to remain orientated throughout the day by making allowance for the daily movement of the sun. This has been demonstrated by experiments in which various species of fishes have been trained to swim at certain angles to the sun at specific times. The angle of movement to the sun's position changes throughout the day, so the fishes must compensate for the azimuth movement of the sun.[12] In other experiments fishes were trained in the northern hemisphere and then flown to the equator and the southern hemisphere, where they continued to change their angle of movement relative to the sun's position during the day, just as they had in the original northern latitude.

Many fishes perform impressive migrations, both in the ocean and fresh water, and are probably able to do so in response to time-compensated solar navigation. The same is true of sea turtles, which may swim for thousands of kilometres to the beaches on which they land and lay their eggs. Not only bats, but terrestrial mammals such as bison and caribou, as well as marine whales, regularly perform long migrations. It is probable that their navigation, at least to some extent, depends upon the possession of biological clocks.

Bi-coordinate bird navigation

Some birds displaced on migration have been able to correct for that displacement and regain the original goal. Such homing ability has been studied extensively by experiments in which birds were removed from their nests. Other data has been obtained from observations on homing pigeons. G. V. T. Matthews summarizes as follows the evidence indicating that birds can actually determine and correct for displacement. Birds have returned to known territory from regions with which they were not familiar so quickly that it would have been impossible for them to have arrived as a result of random wandering. Examples include a Manx shearwater (*Procellaria puffinus*) homing some 4,800 km. (3,000 miles) from Boston, Mass. to Wales in twelve and a half days; a Leach's petrel (*Oceanodroma leucorrhoa*) flying from Sussex, England to Maine in fourteen days; and a Laysan albatross (*Diomedea immutabilis*) homing 5,150 km. (3,200 miles) to Midway Island across the Pacific from Washington State in ten days.[13]

Releases from shorter distances have likewise resulted in homing times so short that they could only have been accomplished by direct flight, while observations made *en route* have confirmed that the birds were on the direct line home. In cases where their tracks were plotted by biotelemetry, birds have made remarkably few deviations. Finally, when homing direction has been observed from the ground, it has been found that birds are lost to sight in the general direction of home within a few minutes of release.

Many suggestions have been made to explain such phenomena, but it is now believed that the navigational process involves comparison of stimuli at the point of release with those remembered at home. Two hypotheses have been put forward. According to the first, the

Plate 1. A very large female camel-spider, with a total body length in excess of 6 cm. Despite their enormous jaws, these desert-dwelling arachnids are nocturnal in habit because they are vulnerable to vertebrate predators.

Plate 2. Cicadas show a pronounced diurnal periodicity of sound production, and their singing is strictly controlled by the temperature.

Plate 3. The diamond-back rattlesnake of Arizona is nocturnally active and seeks shade from the midday heat.

Plate 4. The Indian cobra follows a strict rhythm of activity. Although deaf to air-borne sounds – it sways in time to the movements of the snake charmer's flute to which it responds visually – the cobra is sensitive to vibrations of the ground.

Plate 5. This baby Nile crocodile already displays the rhythmic patterns of behaviour by which the adults regulate their body temperature. Here it is gaping and thereby cooling itself by evaporation of moisture from the mouth.

Plate 6. The daily life of the hippopotamus displays a leisurely rhythm. The daytime is passed resting in water and the animal emerges at dusk to graze on land when the heat of the day is over.

Plate 7. These Tunisian toads, photographed by flashlight at the entrance to their hole, spend the day in cool damp surroundings a metre or more below ground.

Plate 8. The eye of the cheetah, like that of many other carnivores, is adapted for vision both during the day and at night. The retina possesses both rods and cones, and a tapetum which reflects the light.

Plate 9. Barnacles and dog-whelks are inactive at low tide, but they resume feeding as soon as they are covered by the waves.

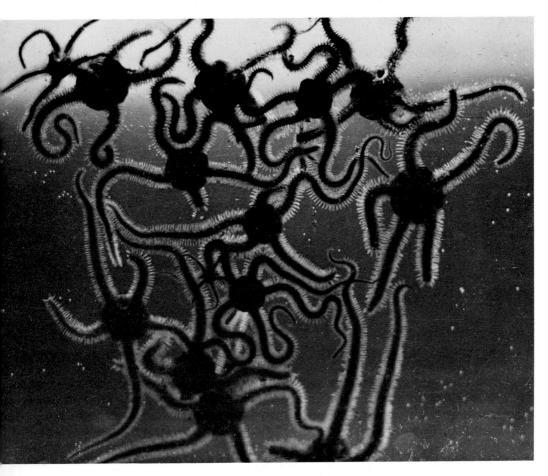

Plate 10. Even below low-tide marks, a lunar rhythm is reflected in the movements of many littoral animals such as the brittle stars photographed here in an aquarium.

Plate 11. The coming of spring is heralded by the profuse blossoms of the wild cherry tree.

Plate 12. Flamingos migrating between their daytime feeding area and their roosting site for the night.

bird observes the movement of the sun along a small portion of the arc and then, by extrapolation, estimates the highest point and the time at which it would be reached. The sun's altitude at noon is inversely proportional to the observer's latitude, while longitudinal displacement is indicated by the difference between the time when the sun is at its highest point in the sky and the time of noon at the bird's home.[13] The second hypothesis suggests that the bird observes the slope of the segment of the sun's arc and compares this with that which would be expected at the same time at home.[14] Both methods of navigation would require very fine analysis of the sun's movement and the comparison of observed with remembered values. The first, bi-coordinate hypothesis requires extrapolation, but reduces the memory requirement to a single point, while the second requires memory of the whole of the sun's arc at the bird's home. Both are conceivable, but little of the experimental evidence upon which the first is based has been confirmed by subsequent workers. Each requires a biological clock of extraordinary accuracy. A circadian clock, however constant the length of its period, would soon drift out of phase with solar time unless its period was of exactly twenty-four hours. Moreover, a clock would be of little use for navigational purposes if it were easily synchronized by changes in the transition from light to dark or vice versa. Presumably birds must either have two clocks, one flexible, one more rigid, or else one clock sufficiently rigid to last through the period of transportation in which the bird does not have access to astronomical information.[15] It seems unlikely that birds can be guided by polarized light during their migration, since direct tests have failed to show any sensitivity to the plane of polarization of light.

Orientation by the moon

Very little seems to be known about lunar orientation in animals, but it is not improbable that many nocturnal species navigate by means of time-compensated moon-compass reactions. Nocturnal orientation has been described in sand-hoppers which tend to move inland at night when they are favoured by the lower temperature and higher atmospheric humidity. It is not clear whether the inland migration is orientated or merely represents casual wandering from the daytime habitat, but the return to the sea is certainly orientated either by the sun, as we have already seen, or by the moon. This is indicated by the fact that the animals are disorientated on moonless nights or when clouds obscure the moon, and their orientation is disturbed during the first and last quarters of the moon. If the moon is hidden, the animals will orientate themselves with respect to the light of an electric torch, by forming an angle with the torch like that assumed with the moon, whatever the azimuth position of the torch.[16, 17]

The quantity of food near the shoreline and the humidity of the atmosphere are probably important environmental factors responsible for the nocturnal migrations of sand-hoppers. An insufficient quantity of food near the sea induces the animals to travel towards the upper part of the beach at night when the humidity is high. At the same time, migration is inhibited by low humidities, while rainfall may elicit migration even during the hours of daylight. Since sand-hoppers feed while they are inland, a quantity of organic matter is transferred every night from the inner parts of the beach back to the shoreline. This may be of importance in fertilizing the moist zones of the beach nearest to the sea.[18]

The woodlouse (*Tylos latreillei*) possesses a diurnal orientation mechanism similar to that of sand-hoppers; it may well also orientate at night by means of a moon-compass reaction. Likewise, it seems probable that various ants (*Formica rufa* and *Monomorium salomonis*) do the same. Workers, marching in a straight line at night, are suddenly disoriented when the moon becomes obscured by clouds, and the assumption of a determined angle of orientation has been demonstrated by means of experiments with mirrors.

Although there is no sharp correlation in birds between the intensity of nocturnal migration and the presence or absence of the moon, many studies of caged birds have shown that the moon has a disruptive influence on migration restlessness. Although little research has yet been carried out on the subject, there is some suggestion that mallards (*Anas platyrhyncha*) may show time-compensated lunar navigation when the sky is overcast with cloud sufficiently thick to obscure the stars, yet thin enough to allow the disc of the moon to be visible.

Time-compensated astral navigation

Under experimental conditions many caged birds have demonstrated the ability to maintain their direction of orientation just as well by night as during the day. Much migration, especially of small songbirds, takes place at night, and the suggestion that the stars might be used as clues for orientation received confirmation when garden warblers (*Sylvia borin*), lesser whitethroats (*S. carruca*) and blackcaps (*S. atricapilla*), which had been hatched and reared in completely enclosed sound-proof chambers in constant light, took up appropriate directions during the migratory season when placed under the artificial stars of a planetarium.

In their oft-quoted research, Franz Sauer and his wife Eleanore claimed that, when the dome of the Olbers Planetarium in Bremen was lighted with diffuse light, the birds would point randomly in many different directions. When a blackcap was shown a simulated spring sky it pointed to the north-east, as it might have been expected to do under natural conditions. Under a simulated autumn sky, on the other hand, it pointed south-west, while a lesser whitethroat, which would have been expected to point south-east, towards the Balkans, did just that.[19]

The time relation of apparent solar movements, due to the rotation of the earth and the difference between the lengths of solar and stellar days, is extremely complicated. It would, therefore, be somewhat surprising if a time-compensation element were found to be involved in solar navigation. The orientation of indigo buntings (*Passerina cyanea*) in a planetarium was not deflected by presenting the birds with skies appropriate to three, six or twelve hours earlier or later.[20] It would appear more likely, therefore, that birds obtain directional information from the patterns of the constellations, much as we determine north from the pointers of the Plough. Only the constellations close to the Pole Star never pass below the horizon, however, and it appears that such circumpolar constellations may be essential to the birds. Others may be eliminated from an artificial sky in a planetarium without affecting the birds' orientation.[15] The claim that the autumn or spring direction of orientation is selected according to whichever non-circumpolar constellations are visible would not appear to be substantiated.[21] The orientation of indigo buntings to the pattern of the stars seems to be determined by the internal state of the birds rather than by seasonal changes in the constellations.[20]

The ability to navigate is clearly of importance to most animals, and especially to migratory species. In many cases, visual landmarks or the magnetic sense are used, and time-compensated celestial navigation is unnecessary. But it is a valuable stand-by in cases of emergency and, when very long distances have to be traversed, it becomes the principle method of navigation.[22] It is no accident that the first accurate man-made chronometers should have been constructed for purposes of navigation, for there are few aspects of the life of man and other animals in which accurate time measurement is so important.[23]

Chapter 7

The Moon and Life

FROM EARLY times it has been thought that in order to ensure good growth sowing and planting should be carried out when the moon was waxing, while reaping and cutting needed a waning moon. There is no evidence in support of these beliefs. No doubt it was assumed that, because the moon was growing, the plants would grow in sympathy. Furthermore, the moon was thought to be associated with water which is clearly needed for plant growth. Perhaps the moon was supposed to be watery because of an imagined connection with dew; this is easily explained because dewfall is greatest on cloudless moonlit nights when heat is rapidly radiated from the earth.[1]

Until the Revolution, French law ordained that trees should be felled only under a waning moon. It was believed that timber for building would not be durable unless it was dry; with a waxing moon it was said to be 'moist' and, therefore, less resistant to the attacks of wood-boring insects. Experiments have been made to test this belief, but they have not supported it. The only real effects of the moon on plants, at present known, are that some flowers which open at night bend towards the moon as others bend towards the sun; certain planktonic algae are most abundant at full moon; clover leaflets close less in bright moonlight than on dark nights; stomata of leaves may open in moonlight; and a few seaweeds show a tidal rhythm of reproduction.

Ancient medicine took periodicities into account and early Western medicine inherited a residue from ancient moon cults and from fertility rituals calibrated to the phase of the moon. Early in the eighteenth century, Richard Mead wrote a *Discourse concerning the Action of the Sun and Moon on Animal Bodies* which is filled with examples such as: 'The Girl, who was of lusty full Habit of Body, continued well for a few days, but was at Full Moon again seized with a most violent Fit, after which, the Disease kept its Periods constant and regular with the tides; She lay always speechless during the whole Time of Flood, and recovered upon the Ebb. . . .'

A belief that the size of edible shell-fish varies with the phases of the moon is found in the literature of classical Greece and Rome, as well as that of the Middle Ages. The same belief is common today in many parts of the world: the amount of edible matter in sea-urchins, molluscs and crabs is said to vary with the phases of the moon, and some gourmets even maintain that the flavour also varies. At Suez, sea-urchins and crabs are said to be 'full' at full moon and 'empty' at new moon; at Alexandria the same is said of mussels and of sea-urchins; and at Nice and Naples, and in Alexandria and Greece, urchins are said to be fattest at the full moon like the oysters at Taranto.[2]

Tidal rhythms

Whereas circadian clocks are valuable in most terrestrial environments, clocks set to the rhythm of the lunar day are chiefly important to littoral forms exposed to tidal action. Since there are two tides during each lunar day of 24.8 hours (the interval between successive moonrises), circalunadian rhythms are necessarily bimodal. Like circadian rhythms, they are apparently important in that they give advance warning of cyclic environmental changes. Circalunadian clocks are illustrated by tidal rhythms of migration in diatoms (*Hantzschia virgata*) and flatworms (*Convoluta roscoffensis*), of locomotory activity and colour change in fiddler crabs and shore crabs (*Carcinus maenas*), by the rhythmic opening and closing of the valves of oysters (*Crassostrea virginica*) and quahogs (*Venus mercenaria*), and in the lunar time-sense of sand-hoppers.[3]

In some cases the adaptive significance of lunar and tidal rhythms has not yet been elucidated. To take an example, various prawns come to the surface of the sea near Bermuda about an hour after sunset at regular intervals throughout the year. These intervals coincide with the time of the new moon, but the numbers of prawns reach maxima on the second night of the lunar month and, again, on the twenty-sixth night. Although these prawns normally feed on certain species of polychaete worm, no correlation has been established between the abundance of worms and the intensity of swarming by the prawns. It therefore appears probable that the activity of the prawns may be inhibited by moonlight, but no functional explanation of this has been hazarded.[4]

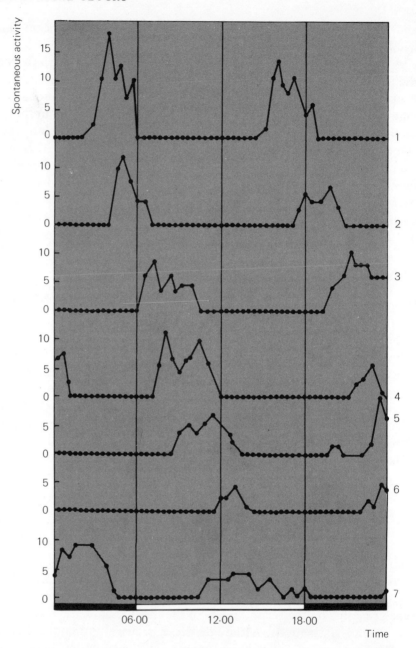

20. A 7-day record of the spontaneous locomotory activity (bursts of running) of a single fiddler crab maintained in constant darkness at 20°C. The peaks of the first day are approximately centred on the time of low tide at the site of collection. Ordinate: spontaneous activity (left) and days of experiment (right). (After J. D. Palmer.)

Under natural conditions, most littoral organisms display both circadian and circalunadian activity rhythms. For instance, the solar rhythm of locomotory activity in crabs is represented, not as an individual peak but as a decrease in the amount of activity at the crest of the daytime tidal peak. A combination of solar-day and lunar-day rhythms is frequently seen in inter-tidal organisms, and it raises the relevant question whether such organisms have a solar-day clock for one rhythmic component and a separate lunar-day clock for the other, or whether a single clock drives both mechanisms. Both hypotheses have been proposed, but neither has yet been proved to be correct.

The tidal cycle on the home shoreline sets the phase of the inhabitant's rhythms. Periodic wetting by inundation is not an important entraining factor for most littoral organisms. Consequently, the cycle is not incorrectly synchronized by rain or spray. Instead, the effective portions of the tidal cycle include one or more of the following: mechanical agitation due to the pounding of the surf, temperature cycles, and the hydrostatic pressure of the water. The last is not generally important for most inter-tidal animals, but it is so in the case of the shore crab.[5] The drop in temperature brought by flood-tides also plays an important role.

Although light–dark cycles have no effect on entrainment, a twenty-four-hour cycle maintains a circalunadian rhythm at strict tidal frequency. Moreover, in rhythms with both daily and tidal components, when the former is shifted by light stimuli, the latter is affected in a nearly identical manner. This might suggest that both rhythms are generated by a single biological clock via specific coupling mechanisms.[6]

Doubtless predation is one of the main selective factors operating on littoral animals to maintain precise tidal rhythmicity. This would explain the significance of rhythmic colour-changes in fiddler crabs and shore crabs, which represent a combination of circadian and circalunadian influences, as well as rhythms of locomotion. Similarly, the nocturnal activity of the ghost crab (*Ocypode ceratophthalmus*) in Mozambique is considerably influenced by tides.[7]

Wading birds often follow the edge of the tide when they are feeding and prey upon the infauna of tidal flats. The significance of this behaviour lies in the fact that food organisms are more accessible near the edge of the water than elsewhere. Sand is more easily penetrated when its water content is high, so that boring for prey is easier. Secondly, rhythmic vertical movements of tidal frequency are common among polychaete worms, amphipod crustaceans and molluscs which burrow deeply when the tide is out.

The rhythm of the polychaete ragworm (*Nereis diversicolor*) is probably endogenous, like the tidal rhythm of the shore crab, but that of a small snail (*Hydrobia ulvae*) is said to be exogenous. Desiccation is apparently of much importance, but the ultimate factor is unknown;[8] perhaps it is the avoidance, as far as possible, of the predatory birds whose feeding area is thus restricted to the water's edge.

For animals such as mussels and barnacles, which are fixed to the substrate, the tidal cycle may resemble day–night transitions, with relatively discrete times of the tidal equivalents of dawn and dusk, but the actual timing of first immersion on any given day can be affected by wave conditions and will, therefore, be less predictable than the dawn whose timing is only slightly affected by cloud-cover. For the animals that migrate up and down sandy beaches with the rising and falling tides, concurrent environmental stimuli would be conspicuously unreliable indicators of the critical phase of the tide.[9]

The survival of tidal migrators, such as sand-hoppers, depends upon endogenous timing which initiates downshore migration at the right tidal phase and prevents their being marooned on the high beach as the tide begins to ebb. Thus many small crustaceans become active when the tide is ebbing and the rhythm is entrained to changes

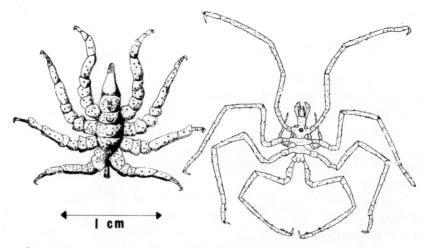

| 1 cm

21. Sea-spiders or pycnogonids avoid being stranded by swimming actively when the tide is ebbing. Two common Atlantic species are illustrated here.

in pressure. The animals are swept landwards by the advancing waves and re-bury themselves at the peak of the upper tidal zone. During the falling tide, however, they continue swimming in a manner that causes them to follow the waves back down the beach and thus avoid being stranded.[10] Similar ebb-tide swimming has also been demonstrated in sea-spiders (*Nymphon gracile*).

Zonation is also maintained by means of an endogenous tidal rhythm of swimming in the intertidal crustacean *Eurydice pulchra*. This predaceous isopod leaves the sand with the incoming tide, feeds actively during high tide, and returns to the sand on the ebb. The animals evidently rely on being washed from the sand by the rising tide, the endogenous component of the rhythm ensuring that they swim for five or six hours before they re-burrow in the sand. The rhythm is synchronized by wave action and reinforced by variations in hydrostatic pressure. The number swimming is also affected by light: fewer animals swim by day since well-fed animals are photo-negative and re-burrow immediately after they emerge into the surf.[11]

Lunar breeding cycles of marine animals

Many marine plants and animals show bi-monthly or monthly lunar reproductive cycles in which all the members of a species within a particular region tend to become sexually active at the same time. This synchronization is essential to the maintenance of the species because it ensures that the gametes are discharged in concentrations sufficiently high to provide a reasonable chance of fertilization taking place.[12] Examples are afforded by the breeding rhythms of palolo worms (*Eunice* spp.), ragworms (*Platynereis dumerilii*) and the queen scallop (*Chlamys opercularis*).

A remarkable correlation is found between the swarming of palolo worms (*Eunice viridis*) in the Pacific and the state of the moon. At the last quarter of the moon in October and November, the posterior portions of the worms, laden with genital products, become detached from the anterior portions, and swim to the surface of the ocean where they shed their eggs and spermatozoa.[13] The main rising occurs at dawn, when great funnels of worms burst to the surface and spread out until the whole area is a wriggling mass of green and brown. The worms provide an annual feast for sharks and other large fishes, which cruise quietly along gulping them in. Clearly isolated worms would stand no chance of survival and it is only by

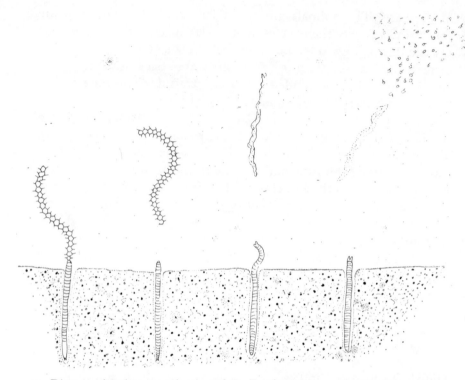

22. Diagram showing reproduction of the Pacific palolo worm. The worms first reverse their positions in their burrows, then the reproductive posterior regions of the body break free and swim towards the surface.

synchronized maturation that a proportion of the worms is able to reproduce.

Another species of palolo worm (*E. fucata*) swarms in the Atlantic, near the Tortugas Islands. Like its relative in the Pacific, the worm lives in great numbers in shallow water, in holes and crevices in old coral, or in honeycombed limestone rock. Maturation of the eggs and sperm is closely adjusted to within one to three days of the last quarter of the moon in late June or early July when, again, the worms commence to swarm just before sunrise.

In contrast, another polychaete ragworm (*P. dumerilii*) performs nuptial dances for a period of an hour or so after sunset on the first and last quarters of the moon at Concarneau in Brittany. At Naples, on the other hand, the same species swarms on the third day after full moon from March to October, with a peak in July and August. No doubt the worms are sensitive both to moonlight and to tidal changes. In the Mediterranean, where tidal movements are slight, breeding is synchronized by the light of the moon whereas in Brittany it is the tidal rhythm which exerts the greatest effect.[14] Again, synchronization enhances the chances of successful reproduction.

Less spectacular is the lunar breeding of oysters (*Ostrea edulis*) most of whose larval swarms occur about ten days after full or new moon throughout the reproductive period which, on the coast of Holland, extends from early June until late August, with a peak as a rule in mid-July. Since swarming takes place about eight days after actual spawning of the eggs into the mantle cavity, it follows that spawning must be at a maximum on the spring tides occurring two days after full and new moon.[14] It is not known exactly what environmental factor synchronizes and controls this remarkable rhythm, but it is probably the rhythmical sequence of differences in water pressure from neap to spring tides. Similarly, there is a physiological rhythm in the queen that causes the development of the reproductive organs to coincide with the full moon of each lunar month during the breeding season. This periodicity is apparently not caused directly by tidal effects, food, or the light of the moon, although one or all of these may serve to entrain the rhythm.[15]

A marine midge (*Clunio marinus*) lays its eggs on masses of seaweed that are accessible only at low water of the semi-monthly spring tides which occur a day or two after full and new moons.[16] The adult female encloses from the pupal state, mates and lays her eggs, within a period of a few hours. The pupal stage lasts for three to five days, however, so the spring tides must be anticipated by this amount when the larvae pupate. Synchronization with the tide is usually achieved by interaction between the circadian rhythm, entrained by the daily cycle of daylight and darkness, and a fortnightly rhythm entrained by moonlight. The full moon is an unreliable entraining agent compared with the new, since the moon is often obscured by clouds, so an endogenous semi-lunar rhythm is really essential for reliable timing.[9]

The grunion (*Leuresthes tenuis*), a small pelagic fish which spawns on the beaches of southern California between late February and early September, also provides a striking example of lunar rhythm. The spawning runs occur only on three or four nights following each full or new moon, and last for a period of from one to three hours immediately after high tide. The eggs mature in about fifteen days and are just ready to hatch when the next spring tides lap them.[14] Quite possibly this rhythm of breeding, synchronized as it is by moonlight, enables the eggs to develop above the water line, where they are in moist sand but safe from predation by fishes and other enemies.

Although few terrestrial animals are known to show lunar breeding rhythms the fact that some do so suggests very strongly that the synchronizer is moonlight itself. In a few instances lunar rhythms of reproduction have been demonstrated among birds and mammals, and it is possible that the precise timing of the breeding seasons of bats, and other inhabitants of the equatorial rain-forest, may be ensured by a lunar rhythm imparting great accuracy to an internal circannual rhythm. This would imply the existence of both circadian and circalunadian clocks.

Moon phase and animal activity on land

The phases of the moon have been shown to affect the activity of at least some nocturnal terrestrial animals.[16] For instance the activity of night-flying insects in England is usually lower at full moon than at new moon, and the activity of aquatic insects is also depressed by moonlight. Feeding behaviour and other activity patterns may have evolved as protective or avoidance mechanisms for escaping the attack of predators that seek their prey visually.[17]

In contrast, adult may-flies (*Povilla adusta*) appear in large numbers in East Africa only at about the date of the full moon. It seems most likely that the influence of the moon induces a rhythm of emergence from the water, and thus affects the sizes of adult winged populations.[18]

Light-trap catches of aquatic insects may be greater at full or new moon, depending on the species concerned. Thus, various species of midges show lunar rhythms of emergence from Lake Victoria; light-trap catches are lowest around full moon and highest at new moon.[19] The swarming of marine midges (*Clunio marinus*), as already mentioned, is likewise synchronized by the state of the moon.[16] It seems probable that such swarming not only facilitates reproduction but also ensures satiation of predators, thus facilitating the survival of at least some of the emerging insects. Anyone who has seen how rapidly a haze of lake-flies over the Nile at dusk can be cleared by insectivorous bats will realize how little chance of survival would be accorded to insects that emerged in small numbers over longer periods of time.

If this argument is accepted it will be seen that provided synchronization of emergence is achieved, it does not matter much whether swarming occurs at full or at new moon. It may, however, be advantageous for different species to swarm at different times of the night, as we saw in the case of doryline ants. In this context, it is significant that one species of midge (*Tanytarsus balteatus*) has a peak of activity just after sunset and another (*Chironomus tetraleucus*) just before dawn.[19]

The activity of a number of terrestrial vertebrates has been shown to be related to the phase of the moon. Increased mating activity occurs during new moon in Javanese frogs (*Rana cancrivora*) and during full moon in a Javanese toad (*Bufo melanostictus*). These conclusions are based on the hormonal condition of the frogs at the time of capture and, in the case of the toads, on the number of pairs found in amplexus. Daily increases in the number of toads (*Bufo biporeatus*) in water were found at Bandung, Java as the moon waxed, but the activity of an American toad (*B. fowleri*) is suppressed during full moon. Moon phase affects changes in rates of activity of another American species (*B. americanus*) which is less active at times of full moon and more active during the new moon, but other factors, especially rain, temperature and relative humidity can mask this effect.[20]

The reasons for the increase in activity at new moon or decrease at full moon are not clear. Toads are insectivorous, and it may be that their prey is more active under low light intensities. Increased availability of food at new moon has also been proposed as an explanation of the lunar rhythm of activity in grasshopper mice (*Onychomys leucogaster*). Moreover, like insects, toads may avoid predation by nocturnal predators, such as skunks and raccoons, by being less active on bright nights. The most plausible hypothesis is that a dual mechanism operates. By hunting just after dusk, and more so on relatively dark nights, toads benefit from maximum insect activity and reduce the chances of random encounters with predators during the long nights.

Lunar periodicity in mammalian reproductive processes

The belief in a connection between human sexual periodicity and the phase of the moon is of very long standing. An exhaustive examination, with data of more than ten thousand menstruations, has, however, disclosed a great variation in the duration of the human cycle, both in the same individual and between different individuals. The average duration is twenty-nine days, but there is a clear inverse correlation between age and the duration of the cycle.[21] In contrast, a ten-year record of services of cows by two Indian buffalo bulls, kept for stud purposes in an agricultural research station, have shown that most of these occur at the time of the new moon, more cows coming into heat on dark than on moonlit nights.[22] Again, in Malayan forest rats, there is a strong tendency for conceptions to be most frequent in the period before full moon. This is true for the nocturnal forest species and, to a less marked degree, for house rats and the rats on an oil-palm estate, but not for day-active forest squirrels.[23]

Daily birth data for the years 1972–3 compiled from the records of the Vancouver General Hospital, on the other hand, give no indication that peaks in the number of deliveries of babies are related to the full moon, and imply that a relation between delivery and the lunar cycle, if indeed one exists, is of minor proportions.[24]

Consideration of these facts suggests that the human menstrual cycle may be the manifestation of a true circalunadian rhythm, but that a civilized, indoor life with artificial lighting prevents most women from being enough influenced by the moon for their rhythms to become properly entrained. Synchronization by means of artificial moonlight could, perhaps, result in greater stability of the cycle.[25]

In our consideration of lunar rhythms and their functions, we have thus come full circle. We began with speculations, and to speculation we return. The influence of the moon on animal life is probably greater than we generally realize. Having freed himself from the tyranny of his own circalunadian clocks, man is slow to appreciate their importance to creatures less fortunate in this respect than himself. A considerable amount of research will be necessary to clarify the situation as far as much of the animal kingdom is concerned.

Chapter 8

Seasonal Adaptations

BIRD MIGRATION has been observed since ancient times. Anacreon, the Greek writer of the late 6th century BC welcomed the return of the swallows in his lyric poetry and assumed correctly that Egypt was one of the birds' winter retreats. The arrival of migratory flocks of quail is a familiar event in Mediterranean countries. According to the *Book of Numbers* one of these, blown off course, saved the Israelites from starvation during their wandering in the wilderness: 'And there went forth a wind from the Lord, and brought quails from the sea, and let them fall by the camp . . .'

An extraordinary idea once prevailed that the swallow passed the cold months at the bottom of a pond or river. Samuel Johnson was merely reflecting popular opinion when he said: 'Swallows certainly sleep all winter. A number of them conglobulate together by flying round and round and then, all in a heap, throw themselves under water and lie in the bed of a river.' An even more extraordinary myth, which existed in many forms, concerns the barnacle goose (*Branta leucopsis*). It was believed that these birds were produced on trees beside the sea, from which they hung, like fruit, attached by their beaks. When mature, they fell into the waters and floated away, while those that fell on the ground died and were lost. No doubt there was confusion between geese and the barnacles, growing on logs of driftwood, whose shape superficially resembles that of the birds.

Yet the truth about bird migration is scarcely less amazing. It has been the subject of many articles and books. Linnaeus wrote an academic thesis on the subject in 1757 and, even today, we are far from fully understanding this extraordinary phenomenon.

Plants and animals have evolved many ways by which they are able to adjust themselves to, avoid and even anticipate seasonal changes on earth. Migration is one of them. In order to become preadapted to the forthcoming season, however, organisms must receive some signal from the environment to trigger their responses. For many centuries it was assumed that seasonal changes in temperature and rainfall caused all the biological phenomena that accompany the changing of the seasons. With the discovery, in 1920, of the photoperiodic control of flowering in plants biologists began to realize that annual changes in day-length could be of even greater biological significance.

Different species of plants and animals are affected in different ways by changing photoperiod. In this chapter we will consider some of the more important effects of changing day-length on animal physiology and behaviour, including seasonal rhythms of reproduction and feeding behaviour; diapause; migration; hibernation; colour-changes and activity patterns. Even metabolism is linked to an animal's photoperiodic responses. In the temperate regions of the world, the survival of an animal from one year to the next is dependent upon its ability to synchronize its activities with the changing seasons, and with those of the other animals and plants with which it is associated.

Synchronization of seasonal rhythms

The reproductive physiology of many plants and animals responds to fluctuations in photoperiod or day-length. This response enables them to breed at the period of the year when food for the young is most plentiful and other environmental conditions are favourable. At the same time, the sexual cycle often responds to other environmental stimuli. Especially in equatorial regions, where changes in photoperiod are slight and occur twice annually, the traditional response to photostimulation is frequently abandoned in favour of other timing devices which ensure that the young are produced at the season most propitious for their survival. The partly endogenous and partly exogenous reproductive cycle of birds, for example, involves successive phases of post-nuptial regeneration (which is reflected in subsequent recovery after sudden loss of breeding functions), acceleration (characterized by the production of sex-hormone and gametogenesis), and culmination (involving ovulation and insemination). The circannual clock is the primary seasonal initiator and, until post-nuptial regeneration is over, male birds and probably also females are not influenced by the external stimuli that would cause the production of sex cells at other periods of the cycle.

After the spontaneous progression from regeneration to

acceleration, the neuro-endocrine machinery comes under the influence of two antagonistic sets of external factors – accelerators and inhibitors. For example, among tropical species, the advent of dry weather often acts as an inhibitor. Many savanna birds habitually breed as soon as the wet season begins, but the acceleration phase is usually under way long before the rains come.[1]

In equatorial regions, the continuous abundance of food and the relative absence of inhibitors often permit the abandonment of a more or less precisely timed annual rhythm, while the innate periodicity of the sooty tern (*Sterna fuscata*) permits reproduction four times every three years on Ascension Island in the Atlantic Ocean. Elsewhere, avian circannual clocks are entrained by external synchronizers at least once every year.[2]

Some organisms possess circannual clocks which operate even when environmental signals are eliminated. Examples are afforded by rhythms of seasonal hibernation in ground squirrels, the migratory rhythms of birds, the yearly shedding of their antlers by male deer, and the moulting of crayfishes. In such ways, animals may become preadapted to forthcoming changes: hibernating species lay down surplus fat before the cold season commences and prepare themselves for forthcoming annual events, such as periods of food shortage and inclement weather. Finally, the circadian clock supplies a guide when environmental signals are weak or absent. For example, birds that spend the northern winter near the equator may need to migrate back to temperate regions for their breeding season.[3]

The rotation of the sun has been shown experimentally to synchronize locomotor activity in finches of various species with the rotation of the earth. At definite times of the year, however, free-running activity may occur and it seems likely that an endogenous annual periodicity may be responsible for the varying effectiveness of the weak *Zeitgeber*.[4] The spectral composition, or colour temperature, of the light, unlike the small amplitude of diurnal changes in light intensity, shows large and regular daily oscillations in high Arctic regions and may therefore also act as a synchronizer of animal activity.[5]

Seasonal changes in the time of activity

The influence of seasonal changes on circadian activity rhythms has not been studied in much detail. It is well known that light, temperature, humidity and other environmental factors that change from season to season affect the gross amount of daily activity in animals. In addition these factors may engender seasonal shifts in the times of daily activities. For instance, field studies on the desert leaf-cutter ant *Acromyrmex versicolor* in Arizona have shown that the ants are nocturnal during the hot summer months, but become diurnally active as autumn progresses.[6] In species of harvester ants that share similar requirements for space and food, competition is avoided by daily and seasonal differences in the time at which foraging takes place.[7] It is probable, therefore, that seasonal variations in circadian rhythms may be related both to physical and biotic factors of the environment.

Some Palaearctic desert beetles, which are day-active in spring, shun the heat during the summer months, when they become crepuscular, or even nocturnal, in habit. This seasonal shift in the nycthemeral activity period has been shown experimentally to be endogenous. Similar results have been obtained with Hemiptera, Diptera, Hymenoptera and other insect orders.[8]

Even in less extreme climates than those of deserts, the times of the daily activities of insects may shift according to the season. Fruit-flies, for instance, have a different pattern of activity in summer from that in early spring or late autumn. In the summer, fruit-flies invariably show a marked peak of activity in the evening, usually a small peak in the morning, and little or no activity in the middle of the day. This behaviour is correlated with light intensity. In the early spring and late autumn when temperatures are usually below 15°C, activity appears to be primarily dependent on temperature, increasing as the temperature rises, while light intensity has only a subsidiary effect.[9]

The flight activity of blow-flies (*Calliphora augur* and *C. stygia*) in Australia is bimodal in hot weather with peaks in the morning and evening. In colder months the curves of activity are unimodal. Seasonal variations have been described in the responses of bark beetles (*Blastophagus piniperda*) to light and temperature. The sine of phototaxis and rate of take-off is correlated with temperature in these insects, and both show a clear seasonal variation. The ecological significance of such changes is concerned with dispersal as early as possible in spring, and with preventing the beetles from leaving their winter retreats in autumn, even if higher temperatures should, in exceptional circumstances, prevail for short periods.[10] Other examples are afforded by the times of emergence of dragonflies and other insects. Alterations in the time of activity are mediated by changes in the intensity of the humidity responses of earwigs, woodlice, millipedes and other arthropods.[8]

Seasonal shifts in the times of peak activity also occur in amphibians, reptiles, birds and mammals. They may be related to thermoregulation, feeding and reproduction, and many are mediated by responses to changing photoperiod. The time of activity of most reptiles of the Sahara changes from summer to winter. Snakes such as *Natrix natrix*, *Spaletosophis diaderma* and *Malpolon moilensis* are nocturnal in summer but day-active in spring, autumn and winter. On the other hand, apart from a few exceptions such as *Scincus scincus* and *Chalcides ocellatus*, lizards do not usually change their activity rhythms according to the season.[11]

When day-length is short, moles (*Talpa europaea*) are active throughout the day and have a second period of activity during the night. As day-length increases, however, this appears as a short rest period in the middle of the day, thereby producing a three-cycle rhythm.[12] In contrast, mole rats (*Tachyoryctes splendens*) show a single peak of activity and only leave the nest between 10.00 and 19.00 hours, while another kind of mole rat (*Heliophobius argenteocinereus*) shows a more dispersed and prolonged activity pattern. These differences can be attributed to the different functions of the nest in the two genera. *Tachyoryctes* has a multipurpose nest while that of *Heliophobius* is used solely for rest. No evidence of aestivation has been found in either genus.[13]

By means of laboratory experiments, it has been shown that a reduction in photoperiod, independent of temperature, is the cue that triggers the intensive period of food storing by flying squirrels (*Glaucomys volans*) that is characteristic of the fall. The storing of food by animals subjected to a long photoperiod (15 hours) was inhibited until the day-length changed. Accelerated diminishing photoperiod resulted in animals showing an accelerated storing response. Thus both the onset and the end of activity follow seasonal changes of photoperiod in nature, whereas the intensity of activity is often correlated with temperature.[14]

We have already seen that the time of day at which feeding and maintenance activities are carried out may be related to inter-specific competition. This has been illustrated by a survey of seasonal activity patterns in twelve species of nocturnal rodents inhabiting a sagebrush community in the southern United States. Differences in the time of daily activity were found between potential competitors, so that a degree of temporal isolation is achieved. Activity was correlated with the time that had elapsed since sunset and the amount of moonlight, while temperature, cloud-cover, wind-speed and direction had little effect except in extreme conditions.[15]

Seasonal effects may influence animals directly, or indirectly through the operation of the circannual clock. Thus nocturnal activity of the masked shrew (*Sorex cinereus*) is increased on cloudy nights and when there is rainfall,[16] while the time of activity of nocturnal rodents such as the bank vole (*Cleithrionomys glareolus*), the field vole (*Microtus arvalis*) and the wood mouse (*Apodemus sylvaticus*), depends on the season.[17] Avoidance of moonlight by kangaroo rats in Arizona also varies seasonally. In October, when the grass seeds on which these rodents live become mature, the optimum strategy is to risk some moonlight activity in return for the benefits of building up a large cache. This reaches a maximum in mid-November when surface food supply has returned to moderate levels. At this time of year the nights are long and it is not worth the added risk of foraging in moonlight, when the kangaroo rats would be more vulnerable to predators. It is only in spring when the nights are shorter, caches begin to dwindle and food becomes scattered and scarce, that foraging in moonlight again becomes necessary.[18]

The seasonal migration of the wildebeest (*Connochaetes taurinus*) in Serengeti is very well documented.[19] Large aggregations of animals, known as armies, are smaller and more restless during the wet season in woodlands than they are on the plain. It appears that while in the bush they are more successful at finding water than green foliage, and most of their movement is necessitated by their search for this.[20]

In the tropics, annual periods of drought or rainfall are often responsible for seasonal breeding and peaks in population numbers. Land snails (*Limicolaria martensiana*) reproduce throughout the year in Uganda, but there are two main times of egg-laying, one in February and the other in July, the two driest months of the year. The ecological significance of this is that the young snails are consequently most abundant at the times of maximum plant growth, when there is least danger of desiccation.[21] Seasonal activity in relation to the time of the annual rains has been described in ticks and other Arachnida, grasshoppers, termites, ants, mosquitoes, beetles and other insects, fishes, amphibians, reptiles, birds and mammals including forest rodents, gerbils, jerboas, antelopes, hippo-potamuses, elephants, and monkeys.[22] Although the cause of seasonal periodicity may not be entirely endogenous, it seems probable that in most species there is an innate rhythm which is kept in adjustment with the annual cycle by means of external stimuli. What these stimuli are and how they function is not always clear, but a selective advantage, achieved by breeding at a particular time of year, can often be inferred.

The daily activities of the desert gazelle (*Gazella dorcas*) consist of fighting, running, walking, feeding, defaecating, chewing the cud, standing, lying, drinking, courtship and so on, all of which are related to the ambient temperature, time of day and season (see page 35).[23] Similarly the different diurnal activities of the African elephant (*Loxodonta africana*)–grazing, browsing, wallowing, resting in shade or sun, walking and taking part in social activities–are related to season, weather conditions and so on.[24, 25]

Parasite rhythms

Annual cycles of infection by parasitic intestinal nematodes have been described in a number of domesticated animals. Such cycles are presumably due to the interplay of many factors, including immunological ones, and are entrained by the return of warm weather in the spring and a fresh supply of susceptible lambs or calves. The eventual biological result is that both parasite and host coexist in equilibrium. The initial multiplication of the parasite is checked after a suitable interval so that it does not wipe out the host while, on the other hand, the immunological defences of the host do not exterminate all the parasites, but leave some survivors to produce new infections. This result may be due to mutual immunity, in which the host can kill new invading parasites but not the adult worms that succeeded in establishing themselves at the primary infection.[26]

Photoperiodism and diapause

In most regions of the world, the physical and biological conditions suitable for growth, development and reproduction prevail only during particular seasons of the year. Some animals avoid inclement periods by migration while others synchronize their activities with favourable seasons, and survive in a state of dormancy during unfavourable ones.

Of the diverse physiological mechanisms controlling dormancy, quiescence and diapause are the most important, The former is an immediate response to adverse conditions, and recovery occurs soon after they have disappeared. For example, insects in a state of cold stupor resume activity when the temperature rises. Diapause, on the other hand, is controlled by a biological clock and depends upon the

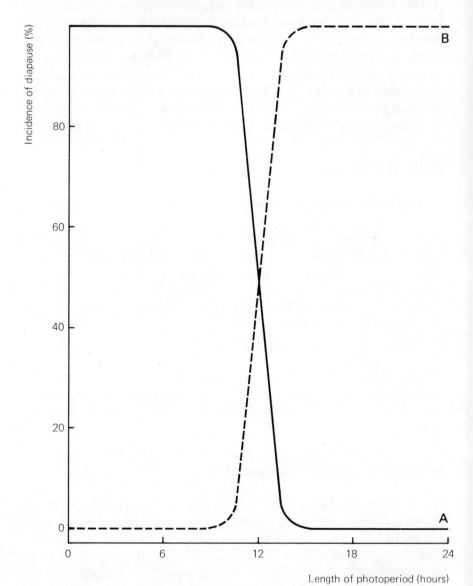

23. A, long-day and B, short-day types of diapause response to the length of photoperiod. (After B. Lofts.)

accurate measurement of day-length. It is induced by environmental changes that take place before the advent of the adverse conditions to which dormancy is an adaptation.

The dormant state of diapause is usually characterized by the temporary failure of growth or reproduction, by reduced metabolism, and often by enhanced resistance to climatic factors such as heat, cold or drought, and by other morphological, physiological and behavioural features. The phenomenon is widespread among living organisms and is an especially striking feature of the biology of insects and mites.[27]

The physiology of diapause has been studied intensively. A genetically determined state of suppressed development, the manifestation of which may be induced by environmental factors, diapause is an important adaptive mechanism for surviving not only periods of unfavourable environmental conditions but also periods during which food is scarce or absent. Indeed, one of the simplest aspects of seasonal timing is concerned with food supply. The natural history of an animal must be synchronized with that of its natural diet. For instance, the larvae and adults of the Colorado potato beetle (*Leptinotarsa decemlineata*) feed on the leaves of potato plants. During the late summer and early autumn, when these wither and turn brown, Colorado beetles cease feeding and burrow into the soil where they hibernate until the following spring. By the time that their period of diapause has ended, potato foliage is again available for them to feed on.[28]

Satisfactory physiological explanation of the linkage between the appreciation of changing photoperiod and the appearance of the resting state is still lacking. An inactive neurosecretory system could well be the prime factor initiating the onset of diapause. In some insects, light appears to act directly on the brain, stimulating neurons close to the cerebral neurosecretory cells and consequently affecting development, although rhythmicity is apparently not controlled directly by hormones.

Many insects show specific patterns of behaviour which are associated with the physiological preparations for diapause. Larval forms seek out suitable hiding places in the soil or under stones or bark, often reversing their customary reactions to gravity and light. Other species construct special cocoons in which they will be protected during their period of dormancy. Overwintering pupae of some species have cryptic colouration designed to camouflage them in the absence of foliage.

Seasonal polymorphism in Arthropoda is often closely associated with diapause. For instance, polymorphism of aphids (*Megoura viciae*) depends upon photoperiod. Long day-lengths lead to the production of parthenogenetic females, whose eggs develop without fertilization, while short day-lengths lead to the production of oviparous aphids whose eggs pass the winter in a state of embryonic diapause.[29] Similar relationships between photoperiod, reproduction and diapause have been reported in various insects, mites (*Tetranychus urticae*) and water-fleas (*Daphnia pulex*).[29]

A classical example of this phenomenon is seen in the butterfly

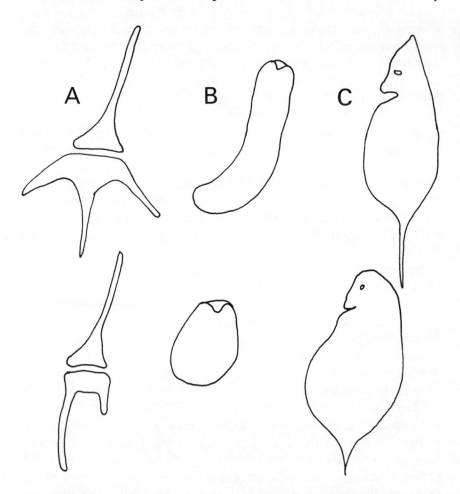

24. Seasonal variation in some planktonic animals from the same lake. A *Ceratium*. B *Asplanchna*. C *Daphnia*. Upper row: summer forms. Lower row: winter forms. (After C. Wesenberg-Lund.)

Araschnia, of which the pale spring form *A. levana* was originally believed to be a separate species, distinct from the dark form (*prorsa*) which appears late in summer. The spring form is produced from diapause pupae, resulting from the short days of the previous autumn, the summer form from non-diapause pupae. By exposing caterpillars to the appropriate photoperiod, either form can be produced experimentally.

Reproductive rhythms are important in determining seasonal peaks of numbers in populations of planktonic animals. Most species undergo periods of dormancy when the active stages disappear and the species is maintained by eggs or developing instars (the form assumed by an insect between two moults) resting in diapause. In some waters of temperate regions, seasonal polymorphism is conspicuous among various species of plankton. During the summer, while rapid multiplication is taking place, there may be a temporary predominance of varieties with well-marked crests, elongated spines and other extensions of the body surface, compared with the more compact forms of the winter generations. This phenomenon occurs in organisms as distantly related as species of *Ceratium* (Dinoflagellata), *Asplanchna* (Rotifera) and *Daphnia* (Crustacea). It is believed that these variations may be related to seasonal changes in the temperature of the water which affects its density and viscosity. Reduced viscosity in summer increases the difficulty of floating and is compensated for by the crests and spines. This type of seasonal polymorphism is limited to shallow lakes with a wide annual range of temperature. In other cases variations may be irregular and their significance less readily understood.

Hibernation

Many vertebrate animals show seasonal breeding cycles, and reproduction is arrested throughout the remainder of the year. Not only is breeding usually controlled by photoperiod – except in amphibians and some other cold-blooded or poikilothermic forms that respond to warmth – but reproduction frequently alternates with migration so that unfavourable seasons are avoided. Reptiles and most mammals that do not migrate, frequently spend the winter in hibernation, a condition brought about mainly in response to low temperature, although day-length also has an effect.

Mammalian hibernation differs from insect diapause in that it is interrupted by short periodic arousals. It enables animals to survive the winter without feeding and with minimum expenditure of food reserves on metabolism. Temperature is regulated at about 1°C above the ambient in the range 5–15°C and arousal can be induced by temperatures above and below this range.

Seasonal colour-change

Many animals respond to environmental challenge by altering the colours of their bodies. Such change may be triggered directly by thermal and other physical stimuli. Many amphibians and reptiles, for example, become lighter in colour when the temperature rises. This lessens their heat load as more of the incoming solar radiation is reflected away. Colour-change may also be associated with sexual and threat displays, but its most important function is related to predation and escape from enemies.

Rhythmic colour-change, whatever its function, may be a direct response to the background colours of the environment. This is seen in many crustaceans, fishes and reptiles. It may also be under the control of the biological clock, as in the case of the fiddler crabs which benefit from preadaptation to the appropriate colour *before* they emerge from their burrows, and not afterwards. The same is true of Arctic birds and mammals that become white in winter so that predator and prey are both less visible against the snowy landscape. When the snow melts in spring they revert to brownish colouring and are difficult to see against the dark soil and sparse vegetation. Change from one kind of cryptic or concealing colouration to another, according to season, is usually synchronized by changing photoperiod. It involves moulting and the appearance of new feathers or hair.

Some species, such as the snowy-owl (*Nyctea scandiaca*), Greenland falcon (*Falco candicans*), and polar bear (*Thalarctos maritimus*), are white all the year round, while others like elk (*Alces alces*), musk-ox (*Ovibos moschatus*), caribou (*Rangifer tarandus*), sable (*Martes* spp.) and glutton or wolverine (*Gulo* spp.) do not become white in winter. Those species that display seasonal colour-change, including willow grouse (*Lagopus lagopus*), ptarmigan (*L. mutus*), Arctic hares (*Lepus arcticus*) and snowshoe rabbits (*L. americanus*), assume white only in the colder parts of their range where the ground is always snow-covered in winter. This makes it difficult for foxes (*Alopex lagopus*) and stoats (*Mustela erminea*) to detect them. These predators, however, themselves become white, which helps them to approach unobserved. Their crypsis is employed in offence, as well as in defence against even larger predators.

Seasonal migration and reproduction in birds

Since W. Rowan published his pioneer work in 1926,[30] it has been recognized that changes in day-length, or photoperiodism, are among the most important factors governing the reproductive cycle in birds, as well as its associated activities including migration. There has been prolonged controversy, however, as to whether cycles of migration and breeding are engendered mainly by endogenous circannual clocks, or whether control is achieved entirely by responses to environmental factors. Neither extreme is tenable; certainly there are cases of animals breeding at other than annual intervals; examples include the sooty tern (*Sterna fuscata*) and Audobon's shearwater (*Puffinus herminieri*) which reproduce every nine months, and the brown booby (*Sula leucosaster*) which breeds every eight months. These are all tropical sea birds which inhabit environments in which there do not appear to be selective pressures favouring breeding at any particular month of the year.

Differences in photoperiodic response mechanism have led to a variety of annual breeding-cycles, and the suggestion has been made that this is based on differences in pattern in the release of two gonadotropic hormones, luteinizing (LH) and follicle-stimulating (FSH). Despite many observations of adrenal–cortical activity as a function of seasonal increase in day-length, no decisive conclusion has been reached as to whether there is a direct causal relationship with photoperiod as there is in the case of gonadal function. Enhanced locomotory activity and nutrition, which are activated seasonally by the long summer days, could easily act as secondary stimulators for adrenal–cortical functions.[31]

As a reaction against accepting too universally the hypothesis of photoperiodicity controlling reproduction, it has been claimed that birds in arid areas have escaped from the routine of regular breeding-cycles. Rainfall and its effects have been postulated as the proximal synchronizer in arid lands, together with the assumption that breeding periods could occur erratically at any season of the year as a consequence of unpredictable and irregular precipitation. It is certainly true that rigid adherence to any particular phase of periodicity would be disastrous in the desert, but only in the arid areas of Australia is the above generalization valid, and even here photoperiod is still involved to some extent, since spring breeding is predominant if rain does fall. The rapid response to rain as a proximate factor initiating nesting in the arid areas of western and central Australia has impressed many observers. The same phenomenon has been recorded in other desert regions but reproduction nevertheless seems not to be entirely freed from the dictates of the photoperiodic clock.[32]

There have been few more extraordinary discoveries about birds in modern times than those which led to the validation, in some measure, of the discredited seventeenth-century belief that birds may, in certain circumstances, tide over periods of stress in a state of torpidity. During an extremely cold spell in January 1913, eight white-throated swifts (*Aeronautes saxatilis*) were taken out of a crevice in the cliffs of Slover mountain, California, where they, with many others, were roosting in a dazed or numb state. During the nineteenth century torpidity had been observed in humming-birds, but it was not until the chance discovery, in 1946, of typical hibernation in the poor-will (*Phalaenoptilus nuttallii*) that widespread interest in the phenomenon was observed.

More recently torpidity as an alternative to migration has been described in the Australian white-backed swallow (*Cheramoeca leucosternum*), crimson chat (*Ephthianura tricolor*), banded whiteface (*Aphelocephala nigricincta*), red-capped robin (*Petroica goodenovii*), white-fronted honeyeater (*Phyledonyris albifrons*) and mistletoe bird (*Dicaeum hirundinaceum*). If it acquired the property of suspended animation, a small bird would survive the cold winter nights of the Australian desert more economically than if it had to ingest extra food to keep warm. Such torpidity is probably a direct response to the cold and not determined by photoperiodism.

Seasonal reproduction in other vertebrates

Experiments have shown that, as in other vertebrates, the annual sexual cycle of fishes is often regulated by photoperiodism. In species that migrate, this phenomenon is often triggered by changes in photoperiod.

Many amphibians undergo seasonal reproductive migrations in which not only do visual and olfactory cues play a role, but gravity also is used for orientation in regions where the topography is abrupt and breeding-pools lie in the bottoms of the valleys. In this connection, it is interesting that the thermal tolerance of frogs may be affected by photoperiod. When leopard frogs (*Rana pipiens*) are subjected to a regime of sixteen hours of light and eight of darkness, their thermal tolerance is significantly higher than it is under short (8:16) or moderate (12:12) photoperiods. It may be of ecological significance that the time of highest temperature tolerance falls during late morning or early afternoon–periods of relatively high environmental temperatures.[33]

There is overwhelming evidence to show that sexual periodicity in numerous species of mammals, as well as in birds, is largely controlled by day-length in temperate regions of the world. The time-sense required to measure photoperiodic changes is of the same order as that required for solar navigation (chapter 6) and, as in the case of diurnal rhythms, increase or decrease in light intensity is likely to give a far more reliable indication of seasonal change throughout the year than are alterations in any other physical factor of the environment. Not all animals show the same sensitivity to increasing or decreasing day-length. In hens and cows, for example, an artificially lengthened day advances the breeding season, while in guinea fowls, which are natives of the tropics, no such effects are shown.

Sexual periodicity in general is often much less pronounced in temperate regions than it is among animals living under tropical or sub-tropical conditions where the seasons are less marked. Even here, however, most species show seasonal rhythms in their activities, although these may not be immediately apparent. Although animals may be found breeding in all months of the year, in any particular species breeding may be restricted to a relatively short period and species that breed continuously tend to show seasonal peaks. Thus the dikdik (*Rhynchotragus kirkii*) shows two peaks of intensive breeding in Tanzania, one at the beginning of the rains in November, and one in April. Elephants, too, tend to show seasonal peaks in breeding, as do some monkeys.

Several domestic animals are partly or very nearly free from the influence of seasonal changes upon their sexual activity, their oestrus cycles being generally accelerated. This has happened almost completely in some domestic cattle and is perhaps partially the result of artificial selection, although warmth, plentiful food and comparatively uniform conditions are probably contributory causes. Nevertheless, oestrus phenomena are much less prominent in winter and may often be absent, although the ovarian cycle continues as 'silent heat', and winter sterility or sub-fertility is common in cattle. Silent heat normally accompanies the first ovulation in the breeding season of the ewe and is reported to occur commonly in the goat. It has been suggested that winter may have been the an-oestrus season in cattle in the wild state and that the ability to breed throughout the year has arisen through artificial selection in the course of domestication. This idea is supported by the fact that, as they get older, cows tend to conceive more readily in June, July and August than at other times of the year.

Seasonal movements have been described in sperm whales (*Physeter catodon*) whose gregarious habits pose interesting problems if stocks are to be rationally exploited. Pairing apparently takes place in the Pacific Ocean south of Durban in summer, and is followed by a northward migration of the females, which overwinter in equatorial waters. These females return again in the spring and calve on their northward migration the following autumn, at about the latitude of Durban. Most of the medium-sized and large bulls swim northwards after the females and smaller males have passed, but some big bulls have been observed associating with the females at the end of the breeding season, and these may be the breeding bulls. Further research will clearly be necessary to elucidate the adaptive function of such migrations of whales.[34]

Although ungulates are better developed and more mobile at birth than the blind naked offspring of many other mammals, the young of most species remain hidden and largely immobile for some days or even weeks after birth. In probably not more than forty out of 187 species do the young follow their mothers from the time they are able to stand. Although wildebeest (*Connochaetes taurinus*) is one of four or five species of antelopes known to have young that follow their mothers, the young of most antelopes are hiders. Each of these strategies includes a number of adaptations that enhances its efficiency. The measures that promote concealment are remarkably similar, even among distantly related species, whereas there are several different strategies for safeguarding follower-young, which are apt to be more exposed to predation even though they are less helpless than are hiders. In some species, such as the rhinoceros, hippopotamus and giraffe, the mother is able to defend her young against lions and spotted hyaenas. Musk-oxen form defensive rings around their calves, but the young of wildebeest, topi, blesbok and caribou have to escape from their predators by flight.

The species whose young follow their mothers and depend upon flight inhabit open country where they enjoy a nomadic or migratory existence. In the case of the wildebeest and other ungulates in the same category, there are two other characteristics – a brief, synchronized peak of reproduction and the formation of large aggregations. Even at the equator, where an extended period of breeding is usual, most wildebeest are born during an interval of about three weeks. Lack of reproductive synchrony between small herds inhabiting different localities renders their calves more vulnerable to predation by spotted hyaenas. In large aggregations there are enough older calves to satisfy the predators and the most vulnerable calves are protected in the confusion created by large numbers.[35]

The advantages of seasonal synchrony are, in many ways, similar to those of synchrony based on daily or lunar rhythms of activity. Despite the diversity of biological phenomena discussed in this book, it has been possible to glimpse unifying concepts of similar physiological mechanisms and of generally applicable adaptational advantages, in rhythms with widely different periodicities.

Conclusion

A FASCINATION with biological clocks, intriguing though they are, should not blind us to the fact that the behaviour and physiology of plants and animals is determined primarily by direct responses to environmental stimuli. Endogenous rhythmicity merely provides the organism with a time-dependent readiness to respond to environmental stimuli in the appropriate manner, and the undoubted advantage of being preadapted to forthcoming events. A selective advantage also accrues to an organism by the possession of an internal timing mechanism when concurrent stimuli are absent as, for example, in a bat cave or down a scorpion's hole, or when environmental synchronizers are confused by inclement weather.

From an ecological point of view, it is clear that internal timing mechanisms have an important influence on the behaviour and responses of plants and animals in the field. In most cases, however, behaviour is determined by rhythmical environmental stimuli and the endogenous clock merely provides the organism with a time-dependent readiness to respond to such stimuli in an appropriate manner. Thus the organism is able to respond to a given stimulus at the correct time, and is not easily misled by minor environmental disturbances.

The biological clock is an adaptable mechanism, and its influence permeates the fields of physiology and ecology. The parameter of time exerts a stabilizing influence on considerations of biological theory, and we may be justified in believing that the importance of biological clocks will come to be appreciated even more as research adds to our knowledge of the basic mysteries of life.

References

CHAPTER I

1 G. G. Luce, *Biological rhythms in psychiatry and medicine* (Maryland 1970).

2 A. Sollberger, *Biological rhythm research* (Amsterdam 1965).

3 F. A. Brown, Jr, 'Response to pervasive geophysical factors and the biological clock problem', *Cold Spring Harb. Symp. Quant. Biol.*, **25** (1960) pp. 57–71.

4 F. A. Brown, Jr, J. W. Hastings and J. D. Palmer, *The biological clock. Two views* (New York 1970).

5 C. S. Pittendrigh and V. G. Bruce, 'Daily rhythms as coupled oscillator systems and their relation to thermoperiodism and photoperiodism', in R. B. Wittrow (ed.), *Photoperiodism and related phenomena in plants and animals* (Washington 1959) pp. 475–505.

6 F. Halberg, M. B. Visscher and J. J. Bittner, 'Relation of visual factors to eosinophil rhythm in mice', *Amer. J. Physiol.*, **179** (1954) pp. 229–35.

7 F. Halberg, E. Halberg, C. P. Barnum and J. J. Bittner, 'Physiologic 24-hour periodicity in human beings and mice, the lighting regimen and daily routine', in R. B. Wittrow (ed.), *Photoperiodism and related phenomena in plants and animals* (Washington 1959) pp. 803–78.

8 J. Aschoff, 'Zeitgeber der tierischen Tagesperiodik', *Naturwiss.*, **41** (1954) pp. 49–56.

9 J. D. Palmer, 'Tidal rhythms: the clock control of the rhythmic physiology of marine organisms', *Biol. Rev.*, **48** (1973) pp. 377–418.

10 J. D. Palmer, *Biological clocks in marine organisms* (New York 1974).

11 C. P. Richter, *Biological clocks in medicine and psychiatry* (Springfield, Ill. 1965).

12 A. Reinberg and J. Ghata, *Biological rhythms*, trans. C. J. Cameron (New York 1964).

13 M. Menaker (ed.), *Biochronometry* (Washington D.C. 1971).

14 E. T. Pengelley (ed.), *Circannual clocks* (New York 1974).

15 C. F. Ehret and E. Trucco, 'Molecular models for the circadian clock, 1. The chronon concept', *J. Theoret. Biol.*, **15** (1967) pp. 240–62.

16 V. G. Bruce, 'Environmental entrainment of circadian rhythms', *Cold Spring Harb. Symp. Quant. Biol.*, **25** (1960) pp. 29–48.

17 C. S. Pittendrigh, 'On temporal organisation in living systems', *Harvey Lectures*, series 56 (New York 1961) pp. 93–125.

18 J. E. Harker, 'Diurnal rhythms in the animal kingdom', *Biol. Rev.*, **33** (1958) pp. 1–52.

19 W. H. Thorpe, *Learning and instinct in animals* (London 1956).

20 W. H. Thorpe, 'Sensitive periods in the learning of animals and men: a study of imprinting with special reference to the induction of cyclic behaviour', in W. H. Thorpe and O. L. Zangwill (eds.), *Current problems in animal behaviour* (Cambridge 1961) pp. 194–224.

21 J. L. Cloudsley-Thompson, *Rhythmic activity in animal physiology and behaviour* (London 1961).

22 J. Aschoff, 'Exogenous and endogenous components in circadian rhythms', *Cold Spring Harb. Symp. Quant. Biol.*, **25** (1960) pp. 11–28.

23 M. F. Bennett, *Living clocks in the animal world* (Springfield, Ill. 1974).

24 J. T. Enright, 'Synchronisation and ranges of entrainment', in J. Aschoff (ed.), *Circadian clocks* (Amsterdam 1965) pp. 112–24.

25 E. Bünning, *The physiological clock*, 3rd edn (London 1973).

26 M. B. Wilkins, 'The influence of temperature and temperature changes on biological clocks', in J. Aschoff (ed.), *Circadian clocks* (Amsterdam 1965) pp. 146–63.

27 E. Bünning, 'Die endogene Tagesrhythmik als Grundlage der photoperiodischen Reaktion', *Ber. dt. bot. Ges.*, **54** (1936) pp. 590–607.

28 R. R. Ward, *The living clocks* (London 1972).

29 P. J. De Coursey, 'Phase control of activity in a rodent', *Cold Spring Harb. Symp. Quant. Biol.*, **25** (1960) pp. 49–56.

30 O. Park, 'Nocturnalism – the development of a problem', *Ecol. Monogr.*, **10** (1940) pp. 485–536.

31 P. S. Corbet, 'The role of rhythms in insect behaviour', in P. T. Haskell (ed.), *Insect Behaviour*, *Symp. Roy. Ent. Soc.*, **3** (1965) pp. 13–28.

32 J. L. Cloudsley-Thompson, 'Biological clocks and their synchronisers', in J. T. Fraser, N. Lawrence and D. Park (eds.), *The study of time* (Proc. 3rd Conf. Internat Soc. Stud. Time, Alpbach 1976; Berlin 1978) pp. 188–215.

33 B. Lofts and R. K. Murton, 'Photoperiodic and physiological adaptations regulating avian breeding cycles and their ecological significance', *J. Zool. Lond.*, **155** (1968) pp. 327–94.

34 J. Aschoff, 'Comparative physiology: diurnal rhythms', *Ann. Rev. Physiol.*, **25** (1963) pp. 581–600.

35 B. M. Sweeney and J. W. Hastings, 'Effects of temperature upon diurnal rhythms', *Cold Spring Harb. Symp. Quant. Biol.*, **25** (1960) pp. 87–104.

36 B. M. Sweeney, *Rhythmic phenomena in plants* (London 1969).

37 J. L. Cloudsley-Thompson, 'Studies in diurnal rhythms. VIII. The

endogenous chronometer in *Gryllus campestris* L. (Orthoptera: Gryllidae)', *J. Insect Physiol.*, **2** (1958) pp. 275–80.

38 A. T. Winfree, 'Unclocklike behaviour of biological clocks', *Nature, Lond.*, **253** (1975) pp. 315–19.

39 V. C. Bruce, 'Cell division rhythms and the circadian clock' in J. Aschoff (ed.), *Circadian clocks* (Amsterdam 1965) pp. 125–38.

40 B. C. Goodwin, 'Temporal order as the origin of spatial order in embryos', in J. T. Fraser, F. C. Habner and G. H. Müller (eds), *The study of time* (Proc. 1st Conf. Internat. Soc. Stud. Time, Oberwolfback 1969; Berlin 1972) pp. 190–9.

41 J. T. Fraser, *Of time, passion, and knowledge. Reflections on the strategy of existence* (New York 1975).

42 R. L. Brahmachary, 'Physiological clocks', *Int. Rev. Cytology*, **21** (1967) pp. 65–89.

43 C. S. Pittendrigh, 'Circadian rhythms and the circadian organisation of living systems', *Cold Spring Harb. Symp. Quant. Biol.*, **25** (1960) pp. 159–84.

44 J. W. Truman, 'Circadian rhythms and physiology with special reference to neuroendocrine processes in insects', *Proc. Int. Symp. Circadian Rhythmicity* (Wageningen 1971) pp. 111–35.

45 J. E. Harker, *The physiology of diurnal rhythms* (Cambridge 1964).

46 J. Brady, 'How are insect circadian rhythms controlled?' *Nature, Lond.*, **223** (1969) pp. 781–4.

47 J. L. Cloudsley-Thompson, 'Time sense of animals', in J. T. Fraser (ed.), *The voices of time* (New York 1966) pp. 296–311.

48 J. R. Baker, 'The evolution of breeding seasons', in G. R. de Beer (ed.), *Evolution. Essays on aspects of evolutionary biology* (London 1938) pp. 161–77.

49 S. D. Beck, *Insect photoperiodism* (New York 1968).

CHAPTER 2

1 J. J. d'O. de Mairan, 'Observation botanique', *Hist. Acad. Roy. Sci. Paris* (1729) p. 35.

2 J. G. Zinn, 'Von dem Schlafe der Pflanzen', *Hamburgischen Magazin,* **22** (1759) pp. 40–50.

3 A. P. de Candolle, *Physiologie végétale* (Paris 1832) p. 859.

4 C. R. Darwin, *On the movements and habits of climbing plants*, 2nd revised edn, (London 1875).

5 E. Bünning and K. Stern, 'Über die tagesperiodischen Bewegungen der Primärblätter von *Phaseolus multiflorus*. II. Die Bewegungen bei Thermokonstanz', *Ber. dtsch. bot. Ges.*, **48** (1930) pp. 227–52.

6 M. B. Jones and T. A. Mansfield, 'Circadian rhythms in plants', *Sci. Progr. Oxf.*, **62** (1975) pp. 103–25.

7 W. W. Garner and H. A. Allard, 'Effect of the relative length of day and

night and other factors of the environment on growth and reproduction in plants', *J. Agric. Res.*, **18** (1920) pp. 553–606.

8 H. E. Street and H. Öpik, *The physiology of flowering plants: their growth and development* (London 1970).

9 K. C. Hamner and J. Bonner, 'Photoperiodism in relation to hormones as factors in floral initiation and development', *Bot. Gaz.*, **100** (1938) pp. 388–431.

10 K. C. Hamner, 'Experimental evidence for the biological clock', in J. T. Fraser (ed.), *The voices of time* (New York 1966) pp. 281–95.

11 F. W. Went, 'Ecological implications of the autonomous 24-hour rhythm in plants', *Ann. New York Acad. Sci.*, **98** (1962) pp. 866–75.

12 M. B. Wilkins, 'Circadian rhythms in plants', in J. N. Mills (ed.), *Biological aspects of circadian rhythms* (London 1973) pp. 235–79.

13 B. Taylor and M. D. R. Jones, 'The circadian rhythm of flight activity in the mosquito *Aedes aegypti* (L.); the phase setting effects of light-on and light-off', *J. Exp. Biol.*, **51** (1969) pp. 59–70.

14 B. M. Sweeney, *Rhythmic phenomena in plants* (London 1969).

15 D. H. Janzen, 'Synchronization of sexual reproduction of trees within the dry season in Central America', *Evolution*, **21** (1967) pp. 620–37.

16 F. B. Salisbury, *The flowering process* (Oxford 1963).

17 J. D. Ligon, 'Reproductive interdependence of piñon jays and piñon pines', *Ecol. Monogr.*, **48** (1978) pp. 111–26.

CHAPTER 3

1 J. L. Cloudsley-Thompson, 'Studies in diurnal rhythms. VI. Bioclimatic observations in Tunisia and their significance in relation to the physiology of the fauna, especially woodlice, centipedes, scorpions and beetles', *Ann. Mag. Nat. Hist.*, 12: **9** (1956) pp. 305–29.

2 K. Schmidt-Nielsen, 'Desert animals', *Physiological problems of heat and water* (Oxford 1964).

3 J. L. Cloudsley-Thompson, *Desert life* (Oxford 1965).

4. C. Darwin, *The formation of vegetable mould through the action of worms* (London 1881).

5 M. Bennett, *Living clocks in the animal world* (Springfield, Ill. 1974).

6 C. L. Ralph, 'Persistent rhythms of activity and O_2 consumption in the earthworms', *Physiol. Zoöl.*, **30** (1957) pp. 41–55.

7 J. Arbit, 'Diurnal cycles and learning in earthworms', *Science*, **126** (1957) pp. 654–5.

8 M. Bennett and D. C. Reinschmidt, 'The diurnal cycle and a difference in reaction times in earthworms', *Zeits. vergl. Physiol.*, **49** (1965) pp. 407–11.

9 M. F. Bennett and D. C. Reinschmidt, 'The diurnal cycle and locomotion in earthworms', *Zeits. vergl. Physiol.*, **51** (1965) pp. 224–6.

10 J. L. Cloudsley-Thompson, 'Adaptive functions of circadian rhythms', *Cold Spring Harb. Symp. Quant. Biol.*, **25** (1960) pp. 345–55.

11 M. S. Ghilarov, 'L'importance du sol dans l'origine et l'evolution des insectes', *Proc. IX Int. Congr. Ent.*, **1** (1958) pp. 443–52.

12 G. S. Fraenkel and D. L. Gunn, *The orientation of animals*, 2nd edn (New York 1961).

13 J. L. Cloudsley-Thompson, 'Studies in diurnal rhythms. II. Changes in the physiological responses of the woodlouse *Oniscus asellus* (L.) to environmental stimuli', *J. Exp. Biol.*, **29** (1952) pp. 295–303.

14 J. L. Cloudsley-Thompson, 'Studies in diurnal rhythms. VI. Bioclimatic observations in Tunisia and their significance in relation to the physiology of the fauna, especially woodlice, centipedes, scorpions and beetles', *Ann. Mag. Nat. Hist.*, 12: **9** (1956) pp. 305–29.

15 J. L. Cloudsley-Thompson, 'Studies in diurnal rhythms. VII. Humidity responses and nocturnal activity in woodlice (Isopoda)', *J. Exp. Biol.*, **33** (1956) pp. 576–82.

16 C. S. Pittendrigh, 'Adaptation, natural selection and behaviour', in A. Roe and G. G. Simpson (eds), *Behaviour and evolution* (New Haven, Conn. 1958) pp. 390–416.

17 V. R. D. Dyson-Hudson, 'The daily activity rhythm of *Drosophila subobscura* and *D. obscura*', *Ecology*, **37** (1956) pp. 562–7.

18 L. R. Taylor and H. Kalmus, 'Dawn and dusk flight of *Drosophila subobscura* Collin', *Nature, Lond.*, **174** (1954) p. 221.

19 H. Dreisig, 'Diurnal activity in the dusky cockroach *Ectobius lapponicus* L. (Blattodea)', *Ent. Scand.*, **2** (1971) pp. 132–8.

20 C. S. Pittendrigh, 'The ecoclimatic divergence of *Anopheles bellator* and *A. homunculus*', *Evolution*, **4** (1950) pp. 43–63.

21 J. L. Cloudsley-Thompson, 'Diurnal rhythm, temperature and water relations of the African toad, *Bufo regularis*', *J. Zool., Lond.*, **152** (1967) pp. 43–54.

22 B. M. Chapman and R. F. Chapman, 'A field study of a population of leopard toads (*Bufo regularis regularis*)', *J. Anim. Ecol.*, **27** (1958) pp. 265–86.

23 J. L. Cloudsley-Thompson, 'Studies in diurnal rhythms. V. Nocturnal ecology and water relations of the British cribellate spiders of the genus *Ciniflo* Bl.', *J. Linn. Soc. (Zool.)*, **43** (1957) pp. 134–52.

24 J. L. Cloudsley-Thompson, 'Terrestrial invertebrates', in G. C. Whitlow (ed.), *Comparative physiology of thermoregulation* (New York 1970) pp. 15–77.

25 J. L. Cloudsley-Thompson, *The temperature and water relations of reptiles* (Watford, Herts 1971).

26 H. B. Cott, 'Scientific results of an enquiry into the ecology and economic status of the Nile crocodile (*Crocodilus niloticus*) in Uganda

and Northern Rhodesia', *Trans. Zool. Soc. Lond.*, **29** (1961) pp. 211–356.

27 N. Kleitman, *Sleep and wakefulness*, revised edn (Chicago 1963).

28 F. Sogaard Andersen, 'Sleep in moths and its dependence on the frequency of stimulation in *Anagasta kuehniella*', *Opusc. Entomol.*, **33** (1968) pp. 15–24.

29 H. Hediger, *Studies of the psychology and behaviour of captive animals in zoos and circuses*, Trans. G. Sireom (London 1955).

30 J. Aschoff, F. Ceresa and F. Halberg (eds), 'Chronobiological aspects of endocrinology', *Chronobiologia*, **1**: suppl. 1 (1974).

CHAPTER 4

1 J. Aschoff, 'Survival value of diurnal rhythms', *Symp. Zool. Soc. Lond.*, **13** (1964) pp. 79–98.

2 F. Halberg, 'Chronobiology', *Ann. Rev. Physiol.*, **31** (1969) pp. 675–725.

3 A. Reinberg, 'Chronopharmacology', in J. N. Mills (ed.), *Biological aspects of circadian rhythms* (London and New York 1973) pp. 121–52.

4 L. E. Scheving, 'The dimension of time in biology and medicine– chronobiology', *Endeavour*, **35** (1976) pp. 66–72.

5 F. Halberg, E. A. Johnson, B. W. Brown and J. J. Bittner, 'Susceptibility rhythm to *E. coli* endotoxin and bioassay', *Proc. Soc. Exp. Biol. N.Y.*, **103** (1960) pp. 142–4.

6 R. H. L. Wilson, E. J. Newman and H. W. Newman, 'Diurnal variation in rate of alcohol metabolism', *J. Appl. Physiol.*, **8** (1956) pp. 556–8.

7 C. R. Johnson, 'Daily variation in the thermal tolerance of *Litorea caerulea* (Anura: Hylidae)', *Comp. Biochem. Physiol.*, **40A** (1971) pp. 1109–11.

8 G. L. Walls, *The vertebrate eye and its adaptive radiation* (New York 1942, reprinted 1967).

9 F. Hawking, 'Circadian rhythms of parasites', in J. N. Mills (ed.), *Biological aspects of circadian rhythms* (London 1973) pp. 153–88.

10 F. Hawking, 'Circadian and other rhythms of parasites', *Adv. Parasitology*, **13** (1975) 123–82.

11 F. Hawking, M. J. Worms and K. Gammage, '24- and 48-hour cycles of malaria parasites in the blood; their purpose, production and control', *Trans. Roy. Soc. Trop. Med. Hyg.*, **62** (1968) pp. 731–60.

12 F. Hawking and J. P. Thurston, 'The periodicity of microfilariae. I. The distribution of microfilariae in the body. II. The explanation of its production', *Trans. Roy. Soc. Trop. Med. Hyg.*, **45** (1951) pp. 307–28, 329–40.

13 C. A. Hopkins, 'Diurnal movement of *Hymenolepis diminuta* in the rat', *Parasitology*, **60** (1970) pp. 255–71.

14   C. P. Read and A. Z. Kilejian, 'Circadian migratory behaviour of a cestode symbiote in the rat host', *J. Parasit.*, **55** (1969) pp. 574–8.

15   C. S. Pittendrigh, 'Adaption, natural selection and behaviour', in A. Roe and G. G. Simpson (eds), *Behaviour and evolution* (New Haven, Conn. 1958) pp. 390–416.

16   S. D. Beck, *Insect photoperiodism* (New York 1968).

17   J. A. Downes, 'The swarming and mating flight of Diptera', *Ann. Rev. Ent.*, **14** (1969) pp. 271–98.

18   G. Williams, 'Seasonal and diurnal activity of Carabidae, with particular reference to *Nebria*, *Notiophilus* and *Feronia*', *J. Anim. Ecol.*, **28** (1959) pp. 309–30.

19   G. Williams, 'Seasonal and diurnal activity of harvestmen (Phalangida) and spiders (Araneida) in contrasted habitats', *J. Anim. Ecol.*, **31** (1962) pp. 23–42.

20   W. Kurtze, 'Synokologische und experimentelle Untersuchungen zur Nachtaktivität von Insekten', *Zool. Jb. Syst.*, **101** (1974) pp. 297–344.

21   C. B. Williams, 'An analysis of four years' captures of insects in a light trap. II. The effect of weather conditions on insect activity; and the estimation and forecasting of changes in the insect population', *Trans. R. Ent. Soc. London*, **90** (1940) pp. 227–306.

22   P. Douwes, 'Activity in *Heodes virgaureae* (Lep., Lycaenidae) in relation to air temperature, solar radiation, and time of day', *Oecologia (Berl.)*, **22** (1976) pp. 287–98.

23   F. Wilson and G. J. Snowball, 'Some effects of temperature on the diurnal periodicity of adult emergence in *Trichopoda pennipes* (Diptera: Tachinidae)', *Austr. J. Zool.*, **7** (1959) pp. 1–6.

24   P. J. M. Greenslade, 'Daily rhythms of locomotor activity in some Carabidae (Coleoptera)', *Ent. Exp. & Appl.*, **6** (1963) pp. 17–80.

25   C. Heckrote, 'Temperature and light effects on the circadian rhythm and locomotory activity of the plains garter snake (*Thamnophis radix hayendi*)', *J. Interdiscipl. Cycle Res.*, **6** (1975) pp. 279–90.

26   G. Chong, H. Heatwole and B. T. Firth, 'Panting thresholds of lizards. II. Diel variations in the panting threshold of *Amphibolurus muricatus*', *Comp. Biochem. Physiol.*, **46**A (1973) pp. 827–9.

27   G. de Coursey and P. J. de Coursey, 'Adaptive aspects of activity rhythms in bats', *Biol. Bull. Woods Hole*, **126** (1964) pp. 14–27.

28   F. M. Sulzman, C. A. Fuller and M. C. Moore-Ede, 'Environmental synchronizers of squirrel monkey circadian rhythms', *J. Appl. Physiol. Respirat. Environm. Exercise Physiol.*, **43** (1977) pp. 795–800.

29   C. A. Fuller, F. M. Sulzman and M. C. Moore-Ede, 'Thermoregulation is impaired in an environment without time cues', *Science*, **199** (1978) pp. 794–6.

30   C. S. Pittendrigh, 'On temporal organisation in living systems', *Harvey Lectures*, series 56 (New York 1961) pp. 93–125.

CHAPTER 5

1 Gaius Plinius Secundus, *Naturalis Historia* (*c.* AD 23–79).
2 N. Kleitman, *Sleep and wakefulness*, revised edn (Chicago 1963).
3 K. Mellanby, 'The daily rhythm of activity in the cockroach *Blatta orientalis*. II. Observations and experiments on a natural infestation', *J. Exp. Biol.*, **17** (1940) pp. 278–85.
4 K. Mellanby, 'The physiology and activity of the bed-bug (*Cimex lectularius* L.) in a natural infestation', *Parasitology*, **31** (1939) pp. 200–11.
5 A. J. Haddow, P. S. Corbet, J. D. Gillett, I. Dirmhirn, T. H. E. Jackson and K. W. Brown, 'Entomological studies from a high tower in Mpanga forest, Uganda', *Trans. Roy. Ent. Soc. Lond.*, **117** (1961) pp. 215–43.
6 P. F. Mattingly, 'Mosquito behaviour in relation to disease eradication programmes', *Ann. Rev. Ent.*, **7** (1962) pp. 419–36.
7 J. L. Cloudsley-Thompson, *The zoology of tropical Africa* (London 1969).
8 J. D. Gillett, *Mosquitos* (London 1971).
9 J. L. Cloudsley-Thompson, 'Adaptive functions of circadian rhythms', *Cold Spring Harb. Symp. Quant. Biol.*, **25** (1960) pp. 345–55.
10 F. T. Abushama, 'Rhythmic activity of the grasshopper *Poecilocerus hieroglyphicus* (Acrididae: Pyrogomorphinae)', *Ent. Exp. & Appl.*, **11** (1968) pp. 341–7.
11 J. L. Cloudsley-Thompson, 'Recent work on the adaptive functions of circadian and seasonal rhythms in animals', *J. Interdiscipl. Cycle Res.*, **1** (1970) pp. 5–19.
12 H. Remmert, 'Tageszeitliche Verzahung der Aktivität verschiedener Organismen', *Oecologia*, **3** (1969) pp. 214–66.
13 K. R. Thomas and R. Thomas, 'Locomotor activity responses in four West Indian fossorial squamates of the genera *Amphisbaena* and *Typhlops* (Reptilia, Lacertilia)', *J. Herpetol.*, **12** (1978) pp. 35–41.
14 O. P. M. Prozesky, 'Notes on the daily drinking patterns of certain bird species', *Sci. Pap. Namib Desert Res. Sta.*, **45** (1969) pp. 69–81.
15 J. L. Kavanau and J. Ramos, 'Influences of light on activity and phasing of carnivores', *Amer. Nat.*, **109** (1975) pp. 391–418.
16 J. L. Kavanau, 'Locomotion and activity phasing of some medium-sized mammals', *J. Mammal.*, **52** (1971) pp. 386–403.
17 C. L. Cheeseman, 'Activity patterns of rodents in Rwenzori National Park, Uganda', *E. Afr. Wildl. J.*, **15** (1977) pp. 281–7.
18 J. L. Kavanau and J. Ramos, 'Twilight and onset and cessation of carnivore activity', *J. Wildl. Manag.*, **36** (1972) pp. 653–7.
19 D. S. Wilson and A. B. Clark, 'Above ground predator defence in the harvester termite *Hodotermes mossambicus* (Hagen)', *J. Ent. Soc. S. Afr.*, **40** (1977) pp. 271–82.

20 P. S. Corbet, 'Temporal patterns of emergence in aquatic insects', *Canad. Ent.*, **96** (1964) pp. 264–79.

21 J. Weir and E. Davison, 'Daily occurrence of African game animals at water holes during dry weather', *Zool. Africana*, **1** (1965) pp. 358–68.

22 O. Park, 'Nocturnalism–the development of a problem', *Ecol. Monogr.*, **10** (1940) pp. 485–536.

23 C. H. Kennedy, 'Evolutionary level in relation to geographic, seasonal and diurnal distribution in insects', *Ecology*, **9** (1928) pp. 367–79.

24 S. C. Crawford, 'The habits and characteristics of nocturnal animals', *Quart. Rev. Biol.*, **9** (1934) pp. 201–14.

25 A. H. Clark, 'Nocturnal animals', *J. Wash. Acad. Sci.*, **4** (1914) pp. 139–42.

26 P. Waser, 'Diurnal and nocturnal strategies of the bushbuck *Tragelaphus scriptus* (Pallas)', *E. Afr. Wildl. J.*, **13** (1975) pp. 49–63.

27 F. A. Lancaster and A. J. Haddow, 'Further studies on the nocturnal activity of Tabanidae in the vicinity of Entebbe, Uganda', *Proc. R. Ent. Soc. Lond.*, (A) **42** (1967) pp. 39–48.

28 A. J. Haddow, I. H. H. Yarrow, G. A. Lancaster and P. S. Corbet, 'Nocturnal flight cycle in the male of African doryline ants (Hymenoptera: Formicidae)', *Proc. R. Ent. Soc. Lond.*, (A) **41** (1966) pp. 103–6.

29 J. Crane, 'Aspects of social behaviour in fiddler crabs, with special reference to *Uca maracoani* (Latreille)', *Zoologica*, **43** (1958) pp. 113–30.

30 F. A. Brown, Jr, 'A unified theory for biological rhythms: rhythmic duplicity and the genesis of "circa" periodisms', in J. Aschoff (ed.), *Circadian clocks* (Amsterdam 1965) pp. 231–61.

31 F. H. Barnwell, 'The role of rhythmic systems in the adaptation of fiddler crabs to the intertidal zone', *Amer. Zoologist*, **8** (1968) pp. 569–83.

32 J. T. Enright, 'Ecological aspects of endogenous rhythmicity', *Ann. Rev. Ecol. Syst.*, **1** (1970) pp. 221–38.

33 J. E. G. Raymont, *Plankton and productivity in the oceans* (Oxford 1963).

34 J. E. Harris and U. K. Wolfe, 'A laboratory study of vertical migration', *Proc. Roy. Soc.*, (B) **144** (1955) pp. 329–53.

35 A. Hardy, *The open sea–its natural history. I. The world of plankton* (London 1956).

36 P. M. David, 'The influence of vertical migration on speciation in the oceanic plankton', *System. Zool.*, **10** (1961) pp. 10–16.

37 J. Mauchline and L. R. Fisher, 'The biology of Euphausiids', in F. S. Russell and C. M. Yonge (eds), *Advances in marine biology*, **7** (London 1969) pp. 1–154.

CHAPTER 6

1   C. R. Ribbands, *The behaviour and social life of honeybees* (London 1953).

2   R. Koltermann, '24-Std-Periodik in der Langzeiterinnerung an Duft-und Farbsignale bei der Honigbiene', *Z. vergl. Physiol.*, **75** (1971) pp. 49–68.

3   I. Beling, 'Über das Zeitgedächtnis der Biene', *Z. vergl. Physiol.*, **9** (1929) pp. 259–338.

4   D. S. Saunders, *Insect clocks* (Oxford 1976).

5   M. J. E. Spitzner, *Ausführliche Beschreibung der Korbbienenzucht im sachsischen Churkreise* (Leipzig 1788).

6   K. von Frisch, *The dancing bees. An account of the life and senses of the honey bees* (London 1954).

7   J. L. Gould, 'The dance-language controversy', *Quart. Rev. Biol.*, **51** (1976) pp. 211–44.

8   F. Santschi, 'Observations et remarques critiques sur le mécanisme de l'orientation chez les fourmis', *Rev. suisse Zool.*, **19** (1911) pp. 303–38.

9   L. Pardi, 'Innate components in the solar orientation of littoral amphipods', *Cold Spring Harb. Symp. Quant. Biol.*, **25** (1960) pp. 395–401.

10 F. Papi and J. Syrjämäki, 'The sun-orientation rhythm of wolf spiders at different latitudes', *Arch. ital. Biol.*, **101** (1963) pp. 59–77.

11 F. Papi and P. Tongiorgi, 'Innate and learned components in the astronomical orientation of wolf spiders', *Ergebn. Biol.*, **26** (1963) pp. 259–80.

12 W. Braemer, 'Versuche zu der im Richtungsfinden der Fische enthaltenen Zeitschätzung', *Verh. deutsch. zool. Ges., Zool. Anz.*, **23** Suppl. (1959) pp. 276–88.

13 G. V. T. Matthews, *Bird navigation*, 2nd edn (Cambridge 1968).

14 C. J. Pennycuick, 'The experimental bases of astronavigation in birds: theoretical considerations', *J. Exp. Biol.*, **37** (1960) pp. 573–93.

15 G. V. T. Matthews, 'Biological clocks and bird migration', in J. N. Mills (ed.), *Biological aspects of circadian rhythms* (London 1973) pp. 281–311.

16 F. Papi, 'Orientation by night: the moon', *Cold Spr. Harb. Symp. Quant. Biol.*, **25** (1960) pp. 475–80.

17 F. Papi and L. Pardi, 'On the lunar orientation of sandhoppers (Amphipoda Talitridae)', *Biol. Bull. Woods Hole*, **124** (1963) pp. 97–105.

18 L. Geppetti and P. Tongiorgi, 'Recerche ecologiche sugli artropodi di una spiaggia sabbiosa del litorale tirrenico. II. Le migrazioni di *Talitrus saltator* (Montagu) (Crustacea–Amphipoda)', *Redia*, **50** (1967) pp. 309–36.

19 E. G. F. Sauer and E. M. Sauer, 'Star navigation of nocturnal

migrating birds', *Cold Spr. Harb. Symp. Quant. Biol.*, **25** (1960) pp. 463–73.

20 S. T. Emlen, 'Bird migration: influence of physiological state upon celestial orientation', *Science*, **165** (1969) pp. 716–18.

21 E. G. F. Sauer, 'Die Sternenorientierung nächtlich ziehander Grasmücken (*Sylvia atricapilla, borin* und *corruca*)', *Z. Tierpsychol.*, **14** (1957) pp. 29–70.

22 K. Schmidt-Koenig, *Migration and homing in animals* (Berlin 1975).

23 J. L. Cloudsley-Thompson, *Animal migration* (London 1978).

CHAPTER 7

1 H. M. Fox, 'The moon and life', *Proc. Roy. Instn. G. B.*, **37**: 163 (1956) pp. 1–12.

2 H. M. Fox, 'Lunar periodicity in reproduction', *Proc. Roy. Soc.*, (B) **95** (1923) pp. 523–50.

3 M. Fingerman, 'Tidal rhythmicity in marine organisms', *Cold Spring Harb. Symp. Quant. Biol.*, **25** (1960) pp. 481–9.

4 J. F. G. Wheeler, 'Further observations on lunar periodicity', *J. Linnean Soc. (Zool.)*, **40** (1937) pp. 325–45.

5 E. Naylor, 'Tidal and diurnal rhythms of locomotor activity in *Carcinus maenas* (L.)', *J. Exp. Biol.*, **35** (1958) pp. 602–10.

6 J. D. Palmer, 'Tidal rhythms: the clock control of the rhythmic physiology of marine organisms', *Biol. Rev.*, **48** (1973) pp. 377–418.

7 R. Barrass, 'The burrows of *Ocypode ceratophthalmus* (Pallas) (Crustacea, Ocypodidae) on a tidal wave beach at Inhaca Island, Moçambique', *J. Anim. Ecol.*, **32** (1963) pp. 73–85.

8 W. J. M. Vader, 'A preliminary investigation into the reactions of the infauna of the tidal flats to tidal fluctuations in water level', *Netherlands J. Sea Res.*, **2** (1964) pp. 189–222.

9 J. T. Enright, 'Ecological aspects of endogenous rhythmicity', *Ann. Rev. Ecology Systematics*, **1** (1970) pp. 221–38.

10 J. T. Enright, 'The tidal rhythm of activity of a sand-beach amphipod', *Z. vergl. Physiol.*, **46** (1962) 276–313.

11 D. A. Jones and E. Naylor, 'The swimming rhythm of the sand beach isopod *Eurydice pulchra*', *J. Exp. Mar. Biol. Ecol.*, **4** (1970) pp. 188–99.

12 J. L. Cloudsley-Thompson, 'Adaptive functions of circadian rhythms', *Cold Spring Harb. Symp. Quant. Biol.*, **25** (1960) pp. 345–55.

13 P. Korringa, 'Relations between the moon and periodicity in the breeding of marine animals', *Ecol. Monogr.*, **17** (1947) pp. 347–81.

14 L. A. Harvey, 'Rhythmic periodicities in animals', *Sci. Prog.*, **48** (1960) pp. 106–13.

15 C. Amirthalingam, 'On lunar periodicity in reproduction of *Pecten opercularis* near Plymouth in 1927–28', *J. Mar. Biol. Assoc.*, **15** (1928) pp. 605–41.

16 H. Caspers, 'Rhythmische Erscheinungen in der Fortpflanzung von *Clunio marinus* (Dipt.: Chiron.) und das Problem der lunaren Periodizität bei Organismen', *Arch. Hydrobiol. Suppl.*, **18** (1951) pp. 415–594.

17 N. H. Anderson, 'Depressant effects of moonlight on activity of aquatic insects', *Nature*, **209** (1967) pp. 319–20.

18 R. Hartland-Rowe, 'The biology of a tropical mayfly, *Povilla adusta* Navas (Ephemeroptera, Polymitareida) with special reference to the lunar rhythm of emergence', *Rev. Zool. Bot. afr.*, **5B** (1958) pp. 185–202.

19 A. Tjønneland, 'Observations on three species of East African Chironomidae (Diptera)', *Årb. Univ. Bergen naturw.* **17** (1958) pp. 1–20.

20 G. J. Fitzgerald and J. R. Bider, 'Influence of moon phase and weather factors on locomotory activity of *Bufo americanus*', *Oikos*, **25** (1974) pp. 338–40.

21 D. L. Gunn, P. M. Jenkin and A. L. Gunn, 'Menstrual periodicity: statistical observations on a large sample of normal cases', *J. Obstet. Gynaecol.*, **44** (1937) pp. 839–79.

22 O. Ramanathan, 'Light and sexual periodicity in Indian buffaloes', *Nature, Lond.*, **130** (1932) pp. 169–70.

23 J. L. Harrison, 'The moonlight effect on rat breeding', *Bull. Raffles Mus.*, **25** (1954) pp. 166–70.

24 B. Schwab, 'Delivery of babies and the full moon', *Canad. Med. Assoc. J.*, **113** (1975) pp. 489–93.

25 E. M. Dewan, 'On the possibility of a perfect rhythm of birth control by periodic light stimulation', *Amer. J. Obstet. Gynecol.*, **99** (1967) pp. 1016–19.

CHAPTER 8

1 B. Lofts and R. K. Murton, 'Photoperiodic and physiological adaptations regulating avian breeding cycles and their ecological significance', *J. Zool. Lond.*, **155** (1968) pp. 327–94.

2 A. J. Marshall, 'Annual periodicity in migration and reproduction of birds', *Cold Spring Harb. Symp. Quant. Biol.*, **25** (1960) pp. 499–505.

3 E. T. Pengelly and S. J. Amundsen, 'Annual biological clocks', *Sci. Amer.*, **224**: 4 (1971) pp. 72–9.

4 F. Krüll, 'The position of the sun is a possible Zeitgeber for Arctic animals', *Oecologia (Berl.)*, **24** (1976) pp. 141–8.

5 F. Krüll, 'Zeitgebers for animals in the continuous daylight of high Arctic summer', *Oecologia (Berl.)*, **24** (1976) pp. 149–57.

6 G. J. Gambra, 'Effects of temperature on the surface activity of the desert leaf-cutter ant, *Acromyrmex versicolor versicolor* (Pergande) (Hymenoptera: Formicidae)', *Amer. Midl. Nat.*, **95** (1976) pp. 485–91.

7 W. G. Whitford, P. Johnson and J. Ramirez, 'Comparative ecology of

the harvester ants *Pogonomyrmex barbatus* (F. Smith) and *Pogonomyrmex rugosus* (Emery)', *Insectes Sociaux*, **23** (1976) pp. 117–32.

8 J. L. Cloudsley-Thompson, 'Seasonal changes in the daily rhythms of animals', *Internat. J. Biometeorol.*, **10** (1966) pp. 119–25.

9 V. R. D. Dyson-Hudson, 'The daily activity rhythm of *Drosophila subobscura* and *D. obscura*', *Ecology*, **37** (1956) pp. 562–7.

10 V. Perttunen, 'The seasonal variation in the responses of scolytids to light and temperature', in R. Chauvin (ed.), *La distribution temporelle des activités animales et humaines*. Quatrième session d'Etudes de l'Union Internationale des Sciences Biologiques, Marseilles, October 1975. (Paris 1977) pp. 129–34.

11 R. Gauthier, 'Ecologie et ethologie des reptiles du Sahara nord-occidental (Région de Beni-Abbès)', *Ann. Mus. roy. Afr. Centr.*, **155** (1967) pp. 1–83.

12 J. A. Woods and A. R. Mead-Briggs, 'The daily cycle of activity in the mole (*Talpa europaea*) and its seasonal changes, as revealed by radioactive monitoring of the nest', *J. Zool. Lond.*, **184** (1978) pp. 563–72.

13 J. U. M. Jarvis, 'Activity patterns in the mole-rats *Tachyoryctes splendens* and *Heliophobius argenteocinereus*', *Zoologica Africana*, **8** (1973) pp. 101–19.

14 I. Muul, 'Behavioral and physiological influences on the distribution of the flying squirrel (*Glaucomys volans*)', *Misc. Publ. Mus. Zool. Univ. Michigan.*, **134** (1968).

15 M. J. O'Farrell, 'Seasonal activity patterns of rodents in a sagebrush community', *J. Mammal.*, **55** (1974) pp. 809–23.

16 G. J. Doucet and J. R. Bider, 'The effects of weather on the activity of the masked shrew', *J. Mammal*, **55** (1974) pp. 348–63.

17 K. Ostermann 'Zur Aktivität heimischer Muriden und Gliriden', *Zool. Jb. allge. Zool. Physiol. Tiere*, **66** (1956) pp. 355–88.

18 R. B. Lockard and D. H. Owing, 'Seasonal variation in moonlight avoidance by bannertail kangaroo rats', *J. Mammal.*, **55** (1974) pp. 189–93.

19. L. Pennycuick, 'Movements of the migratory wildebeest population in the Serengeti area between 1960 and 1973', *E. Afr. Wildl. J.*, **13** (1975) pp. 65–87.

20 J. M. Inglis, 'Wet season movements of individual wildebeests of the Serengeti migratory herd', *E. Afr. Wildl. J.*, **14** (1976) pp. 17–34.

21 D. F. Owen, 'Bimodal occurrence of breeding in an equatorial land snail', *Ecology*, **45** (1964) p. 862.

22 J. L. Cloudsley-Thompson, *The zoology of tropical Africa* (London 1969).

23   L. I. Ghobrial and J. L. Cloudsley-Thompson, 'Daily cycle of activity of the dorcas gazelle in the Sudan', *J. Interdiscipl. Cycle Res.*, **7** (1976) pp. 47–50.

24   P. R. Guy, 'Diurnal activity patterns of elephants in the Sangwa Area, Rhodesia', *E. Afr. Wildl. J.*, **14** (1976) pp. 285–95.

25   J. R. Wyatt and S. K. Eltringham, 'The daily activity of the elephant in the Rwenzori National Park, Uganda', *E. Afr. Wildl. J.*, **12** (1974) pp. 273–89.

26   F. Hawking, 'Circadian and other rhythms of parasites', *Adv. Parasitology*, **13** (1975) pp. 123–82.

27   M. J. Tauber and C. A. Tauber, 'Insect seasonality: diapause maintenance, termination, and postdiapause development', *Ann. Rev. Ent.*, **21** (1976) pp. 81–107.

28   S. D. Beck, *Insect photoperiodism* (New York 1968).

29   A. D. Lees, 'The control of polymorphism in aphids', *Adv. Insect Physiology*, **3** (1966) pp. 207–77.

30   W. Rowan, 'On photoperiodism, reproductive periodicity and the annual migration of birds and certain fishes', *Proc. Boston Soc. Nat. Hist.*, **38** (1926) pp. 147–89.

31   G. V. T. Matthews, 'Biological clocks and bird migration', in J. N. Mills (ed.), *Biological aspects of circadian rhythms* (London 1973) pp. 281–311.

32   D. L. Serventy, 'Biology of desert birds', in D. S. Farner, J. R. King and K. C. Parkes (eds), *Avian biology* (New York 1971) pp. 287–339.

33   J. J. Mahoney and V. H. Hutchison, 'Photoperiod acclimation and 24-hour variations in the critical thermal maxima of a tropical and a temperate frog', *Oecologia (Berl.)*, **2** (1969) pp. 143–61.

34   R. Gambell, 'Seasonal movements of sperm whales', *Symp. Zool. Soc. Lond.*, **19** (1967) pp. 237–54.

35   R. D. Estes 'The significance of breeding synchrony in the wildebeest', *E. Afr. Wildl. J.*, **14** (1976) pp. 135–52.

Glossary

Amplitude Intensity of a rhythm: difference between the maximum and minimum of an oscillation.

Biological clock Physiological mechanism driving a biological rhythm.

Biological rhythm Regularly repeated fluctuation in some biological activity or process.

Circadian rhythm Oscillation with a natural period of about twenty-four hours.

Circalunadian rhythm Oscillation with a natural period of about a lunar day (24.8 hours).

Circalunar rhythm Oscillation with a natural period of about a lunar month. (29.5 solar days).

Circannual rhythm Oscillation with a natural period of about a year.

Compensation point Conditions of light intensity under which the rates of photosynthesis and respiration are equal, and there is neither absorption nor production of carbon dioxide or oxygen.

Crepuscular Of twilight: appearing active at dusk and dawn.

Cycle Pattern of events repeated once or more times in the life of an organism.

Diapause A state of suspended animation in which physiological processes are much reduced.

Diel Occupying twenty-four hours: daily.

Diurnal Of the day: not nocturnal.

Endogenous rhythm Rhythm persisting under constant conditions, i.e. self-sustained.

Entrainment Synchronization of a biological rhythm or clock by means of environmental factors or synchronizers.

Exogenous rhythm Rhythmic response to rhythmically changing environmental stimuli.

Frequency The reciprocal of period: the number of oscillations in unit time.

Free-running Of a rhythm, one that is not entrained and shows its natural period.

Frequency demultiplication Entrainment of an oscillation to show periods which are a multiple of the entraining cycles.

Lunar month Interval between new moons: about 29.5 solar days.

Nycthemeral Appertaining to the cycle of daylight and darkness.

Oscillation Rhythmically repeated cycle of events.

Period Time after which a definite phase of an oscillation occurs again.

Phase Instantaneous state of an oscillation within a period. Changing the phase of a rhythm is called *re-setting* the rhythm, by analogy with moving the hands of a clock.

Photoperiod Length of daylight in each twenty-four hours.

Polymorphism Appearance of more than one form within the same species.

Synchronizer Factor by which an endogenous rhythm or self-sustained oscillation is entrained: *Zeitgeber*.

Tidal rhythm Rhythm of marine animals correlated with the movements of the tide and therefore, when endogenous, circalunadian.

Transients Intermediate cycles that appear during change of phase between two steady states.

Zeitgeber Synchronizer.

Index